Transcendental Numbers

M. Ram Murty • Purusottam Rath

Transcendental Numbers

 Springer

M. Ram Murty
Department of Mathematics and Statistics
Queen's University
Kingston, Ontario, Canada

Purusottam Rath
Chennai Mathematical Institute
Siruseri, Tamil Nadu, India

ISBN 978-1-4939-0831-8   ISBN 978-1-4939-0832-5 (eBook)
DOI 10.1007/978-1-4939-0832-5
Springer New York Heidelberg Dordrecht London

Library of Congress Control Number: 2014936970

Mathematics Subject Classification: 11J81, 11J85, 11J86, 11J87, 11J89, 11J91

Printed on acid-free paper

Springer is part of Springer Science+Business Media (www.springer.com)

My vast transcendence holds the cosmic whirl;
I am hid in it as in the sea a pearl.

Sri Aurobindo, The Indwelling Universal, Collected Poems.

Preface

This book grew out of lectures given by the first author at Queen's University during 2006 and lectures by the second author at the Chennai Mathematical Institute during 2008. These constitute the first 18 chapters of the book intended to be an introductory course aimed at senior undergraduates and first-year graduate students. The primary goal of these chapters is to give a quick introduction to some of the beautiful theorems about transcendental numbers. We begin with some earliest transcendence theorems and thereafter move to the Schneider–Lang theorem. This requires some rudimentary background knowledge in complex analysis, more precisely the connection between the growth of an analytic function and the distribution of its zeros. Since this constitutes an essential ingredient of many of the transcendence results, we discuss the relevant features in Chap. 5. We also require some familiarity with elementary algebraic number theory. But we have tried our best to recall the required notions as and when we require them. Having proved the Schneider–Lang theorem, we introduce some of the accessible and essential features of the theory of elliptic curves and elliptic functions so that the reader can appreciate the beauty of the primary applications. Thus Chaps. 1–18 essentially comprise the material for an introductory course.

The second part of the book, namely Chaps. 19–28, are additional topics requiring more maturity. They grew out of seminar lectures given by both authors at Queen's University and the Institute of Mathematical Sciences in Chennai, India. A major part of these chapters treats the theorem of Baker on linear independence of linear forms in logarithms of algebraic numbers. We present a proof of Baker's theorem following the works of Bertrand and Masser. Thereafter, we briefly describe some of the applications of Baker's theorem, for instance to the Gauss class number problem for imaginary quadratic fields. In Chap. 21, we discuss Schanuel's conjecture which is one of the central conjectures in this subject. We devote this chapter to derive a number of consequences of this conjecture.

From Chaps. 22 to 26, we concentrate on some recent applications of Baker's theorem to the transcendence of special values of L-functions. These L-functions arise from various arithmetic and analytic contexts. To begin, we give a detailed treatment of the result of Baker, Birch and Wirsing. This is perhaps the first instance when transcendental techniques are employed to address the delicate issue of non-vanishing of a Dirichlet series at special points. In Chap. 25, we specialise to questions of linear independence of special values of Dirichlet L-functions. In Chap. 26, we consider analogous questions for class group L-functions.

In Chap. 27, we focus on applications of Schneider's theorem and Nesterenko's theorem to special values of modular forms. These modular forms are a rich source of transcendental functions and hence potential candidates to generate transcendental numbers (hopefully "new"). Of course, one can ask about the possibility of applying transcendence tools not just to modular forms but also to their L-functions. But this will force us to embark upon a different journey which we do not undertake here.

Finally, the last chapter is intended to give the reader an introduction to the emerging theory of periods and multiple zeta values. This is not meant to be an exhaustive account, but rather an invitation to the reader to take up further study of these elusive objects. This chapter is essentially self-contained and can be read independent of the other chapters.

To summarise, we hope that the first part of this book would be suitable for undergraduates and graduate students as well as non-experts to gain entry into the arcane topic of transcendental numbers. The last ten chapters would be of interest to the researchers keen in pursuing the interrelation between special values of L-functions and transcendence.

To facilitate practical mastery, we have included in each chapter basic exercises that would be helpful to the beginning student.

Acknowledgements

We thank the Institute of Mathematical Sciences and Chennai Mathematical Institute, India for their kind hospitality which made this collaborative work possible. We are also thankful to Sanoli Gun, Tapas Chatterjee, R. Thangadurai, Biswajyoti Saha and Ekata Saha for going through several parts of the book. A large part of the later chapters were written up when the second author was visiting the Department of Mathematics, University of Orsay under an Arcus grant and the Abdus Salam International Centre for Theoretical Physics (ICTP), Trieste under the Regular associate scheme. He is grateful to both these institutes for a wonderful working environment.

Kingston, ON   M. Ram Murty
Siruseri, Tamil Nadu   Purusottam Rath
November 2013

Contents

Notations and Basic Definitions

We denote by $\mathbb{N}$ the set of non-negative integers, by $\mathbb{Z}$ the ring of rational integers and by $\mathbb{Q}, \mathbb{R}, \mathbb{C}$ the fields of rational numbers, real numbers and complex numbers, respectively. A number field K is a finite extension of $\mathbb{Q}$ contained in $\mathbb{C}$. $\overline{\mathbb{Q}}$ denotes the algebraic closure of $\mathbb{Q}$ in $\mathbb{C}$.

Given an integral domain R containing a field k, a collection of elements $\alpha_1, \ldots, \alpha_n$ in R is called algebraically dependent over k if there exists a non-zero polynomial $P \in k[x_1, \ldots, x_n]$ such that $P(\alpha_1, \ldots, \alpha_n) = 0$. Otherwise, the elements are called algebraically independent over k. The transcendence degree of a field F over k is the cardinality of a maximal algebraically independent subset of F over k. Since the transcendence degree over k or its any algebraic extension is the same, the notion of algebraic independence of complex numbers over $\mathbb{Q}$ or $\overline{\mathbb{Q}}$ is the same. Thus we simply speak of algebraically independent or dependent complex numbers.

We shall also be concerned with algebraic independence of functions. A meromorphic function on $\mathbb{C}$ is said to be transcendental if it is transcendental over the field of rational functions $\mathbb{C}(z)$. A collection of meromorphic functions $f_1, \ldots, f_n$ on $\mathbb{C}$ is said to be algebraically independent over $\mathbb{C}$ if for any non-zero polynomial $P \in \mathbb{C}[x_1, \ldots, x_n]$, the function $P(f_1, \ldots, f_n)$ is not the zero function. Otherwise, the functions are called algebraically dependent. Thus a function f is transcendental if f and the identity function $I(z) = z$ are algebraically independent. Most of the time, we shall simply write functions to be algebraically independent, the implicit assumption being that the independence is over $\mathbb{C}$.

This notion can be extended to functions in several variables. A collection of entire functions $f_1, \ldots, f_n$ on $\mathbb{C}^d$ is said to be algebraically independent over $\mathbb{C}$ if for any non-zero polynomial $P \in \mathbb{C}[x_1, \ldots, x_n]$, the function $P(f_1, \ldots, f_n)$ is not the zero function. Otherwise, the functions are called algebraically dependent.

We write $f(x) = O(g(x))$ or equivalently $f(x) \ll g(x)$ when there exists a constant C such that $|f(x)| \leq Cg(x)$ for all values of x under consideration.

Throughout for $\alpha \neq 0$, we define α^β to be equal to $e^{\beta \log \alpha}$ where we interpret $\log \alpha$ as the principal value of the logarithm with argument in $(-\pi, \pi]$.

However, we shall be frequently working with the set of logarithms of non-zero algebraic numbers and in this set we allow all possible values of log. This forms a $\mathbb{Q}$-vector space. It is more convenient to realise this as being equal to the set $\exp^{-1}(\overline{\mathbb{Q}}^{\times})$, where exp is the familiar exponential map. This description has an analogous manifestation in the elliptic set-up which we shall come across in the later chapters.

Chapter 1

Liouville's Theorem

A complex number α is said to be an *algebraic number* if there is a non-zero polynomial $f(x) \in \mathbb{Q}[x]$ such that $f(\alpha) = 0$. Given an algebraic number α, there exists a unique irreducible monic polynomial $P(x) \in \mathbb{Q}[x]$ such that $P(\alpha) = 0$. This is called the *minimal polynomial* of α. The set of all algebraic numbers denoted by $\overline{\mathbb{Q}}$ is a subfield of the field of complex numbers. A complex number which is not algebraic is said to be *transcendental*.

An algebraic number α is said to be an *algebraic integer* if it is a root of a monic polynomial in $\mathbb{Z}[x]$. It is not difficult to see that the minimal polynomial of an algebraic integer has integer coefficients.

An algebraic number α is said to be of *degree* n if its minimal polynomial $P(x)$ has degree n. Equivalently, $\mathbb{Q}(\alpha)$ is a finite extension of $\mathbb{Q}$ of degree n.

In 1853, Liouville proved a fundamental theorem concerning approximations of algebraic numbers by rational numbers. This theorem enabled him to construct explicitly some transcendental numbers.

Theorem 1.1 (Liouville) *Given a real algebraic number α of degree $n > 1$, there is a positive constant $c = c(\alpha)$ such that for all rational numbers p/q with $(p, q) = 1$, $q > 0$, we have*

$$\left| \alpha - \frac{p}{q} \right| > \frac{c(\alpha)}{q^n}.$$

Proof. Let $P(x)$ be the minimal polynomial of α. By clearing the denominators of the coefficients of $P(x)$, we can get a polynomial of degree n with integer coefficients which is irreducible in $\mathbb{Z}[x]$ and has positive leading term. Let

$$f(x) = a_n x^n + a_{n-1} x^{n-1} + \cdots + a_1 x + a_0 \in \mathbb{Z}[x]$$

M.R. Murty and P. Rath, *Transcendental Numbers*, DOI 10.1007/978-1-4939-0832-5_1,
© Springer Science+Business Media New York 2014

be this polynomial. We sometimes refer this as the minimal polynomial of α over $\mathbb{Z}$. Then

$$|f(\alpha) - f(p/q)| = |f(p/q)| = \left| \frac{a_n p^n + a_{n-1} p^{n-1} q + \cdots + a_0 q^n}{q^n} \right| \geq \frac{1}{q^n}.$$

If $\alpha = \alpha_1, \ldots, \alpha_n$ are the roots of f, let M be the maximum of the values $|\alpha_i|$. If $|p/q|$ is greater than $2M$, then

$$\left| \frac{p}{q} - \alpha \right| \geq M \geq \frac{M}{q^n}.$$

If $|p/q| \leq 2M$, then

$$\left| \alpha_i - \frac{p}{q} \right| \leq 3M$$

so that

$$\left| \alpha - \frac{p}{q} \right| \geq \frac{1}{|a_n| q^n \prod_{j=2}^{n} |\alpha_j - p/q|} \geq \frac{1}{|a_n| (3M)^{n-1} q^n}.$$

Thus choosing

$$c(\alpha) = \min \left(M, \frac{1}{|a_n| (3M)^{n-1}} \right),$$

we have the theorem. $\square$

Note that the constant $c(\alpha)$ can be explicitly computed once the roots of the minimal polynomial of α are given to us. Also it is not difficult to extend this theorem to complex algebraic numbers of degree n. A multivariable generalisation of this idea is suggested in Exercise 2.

Using this theorem, Liouville proved that the number

$$\sum_{n=0}^{\infty} \frac{1}{10^{n!}}$$

is transcendental. Indeed, suppose not and call the sum α. Consider the partial sums

$$\frac{p_k}{q_k} := \sum_{n=0}^{k} \frac{1}{10^{n!}}.$$

Then it is easily seen that

$$\left| \alpha - \frac{p_k}{q_k} \right| < \frac{c}{10^{(k+1)!}}$$

for some constant $c > 0$. If α were algebraic of degree m say, then by Liouville's theorem, the left-hand side would be greater than $c(\alpha)/10^{k!m}$ and for k sufficiently large, this is a contradiction.

Numbers of the above type are examples of what are called *Liouville numbers*. More precisely, a real number β is called a Liouville number if for any non-negative real number v, the inequality

$$0 < \left|\beta - \frac{p}{q}\right| < \frac{1}{q^v}$$

has infinitely many solutions $p/q \in \mathbb{Q}$.

In 1909, Axel Thue improved Liouville's inequality for algebraic numbers having degree at least 3. More precisely, he proved that if α is algebraic of degree $n > 2$, then for any $\epsilon > 0$, there exists a positive constant $c = c(\alpha, \epsilon)$ such that for all rational numbers p/q with $q > 0$ and $(p, q) = 1$, we have

$$\left|\alpha - \frac{p}{q}\right| > \frac{c}{q^{n/2+1+\epsilon}}.$$

Such a theorem has immediate Diophantine applications. To see this, let $f(x, y)$ be an irreducible binary form of degree $n \geq 3$ with integer coefficients. Then Thue's theorem implies that the equation

$$f(x, y) = M$$

for any fixed non-zero integer M has only finitely many integer solutions. Indeed, we may write

$$f(x, y) = a_n \prod_{i=1}^{n} (x - \alpha_i y) = M$$

so that each α_i is an algebraic number of degree ≥ 2. Suppose that there are infinitely many solutions (x_m, y_m). Without loss of generality, we may suppose that for infinitely many m, we have

$$\left|\frac{x_m}{y_m} - \alpha_1\right| \leq \left|\frac{x_m}{y_m} - \alpha_i\right| \quad \text{for} \quad i = 2, \ldots, n.$$

Further by the triangle inequality,

$$\left|\frac{x_m}{y_m} - \alpha_i\right| \geq \frac{1}{2}\left(\left|\frac{x_m}{y_m} - \alpha_i\right| + \left|\frac{x_m}{y_m} - \alpha_1\right|\right) \geq \frac{1}{2}|\alpha_i - \alpha_1|, \quad \text{for } i = 2, \ldots, n.$$

Thus

$$|M| = |f(x_m, y_m)| = |a_n||y_m^n|\left|\frac{x_m}{y_m} - \alpha_1\right| \cdots \left|\frac{x_m}{y_m} - \alpha_n\right| \gg |y_m^n|\left|\frac{x_m}{y_m} - \alpha_1\right|.$$

By Thue's theorem, we obtain

$$\frac{|M|}{|y_m|^n} \gg \frac{1}{|y_m|^{n/2+1+\epsilon}}$$

which is a contradiction for $|y_m|$ sufficiently large.

Thue's theorem was subsequently improved, first by Siegel to an exponent $2\sqrt{n}+\epsilon$ and then by Dyson and Gelfond independently to an exponent $\sqrt{2n}+\epsilon$. Finally, Roth in 1955 proved that if α is algebraic of degree > 1, then for any $\epsilon > 0$, there is a constant $c(\alpha, \epsilon) > 0$ such that for all rational numbers p/q with $q > 0$ and $(p, q) = 1$, we have

$$\left| \alpha - \frac{p}{q} \right| \geq \frac{c(\alpha, \epsilon)}{q^{2+\epsilon}}.$$

In view of the classical theory of continued fractions, this result is essentially best possible.

We end with the following theorem about algebraic independence of certain Liouville numbers proved by Adams [1]. The proof involves a clever modification of Liouville's original idea.

Recall that complex numbers $\alpha_1, \ldots, \alpha_n$ are algebraically dependent if there exists a non-zero polynomial $P(x_1, \ldots, x_n)$ in n variables with rational coefficients such that $P(\alpha_1, \ldots, \alpha_n) = 0$. Otherwise, they are called algebraically independent. We have the following theorem:

Theorem 1.2 (Adams) *Let p and q be two relatively prime natural numbers greater than 1. Then the two Liouville numbers*

$$\alpha = \sum_{n=1}^{\infty} \frac{1}{p^{n!}} \quad \text{and} \quad \beta = \sum_{n=1}^{\infty} \frac{1}{q^{n!}}$$

are algebraically independent.

Proof. Assume the contrary and let $f(x, y)$ be a non-zero polynomial with integer coefficients such that $f(\alpha, \beta) = 0$. Suppose that $p > q$. We consider the following sequences of rational numbers

$$R_N = \sum_{n=1}^{N} \frac{1}{p^{n!}} = \frac{r_N}{p^{N!}} \quad \text{and} \quad S_N = \sum_{n=1}^{N} \frac{1}{q^{n!}} = \frac{s_N}{q^{N!}}.$$

We first note that there are infinitely many N such that $f(R_N, S_N) \neq 0$. If not, then for all N sufficiently large, we have

$$f \left(\sum_{n=1}^{N} \frac{1}{p^{n!}}, \sum_{n=1}^{N} \frac{1}{q^{n!}} \right) = 0. \qquad (1.1)$$

Further,

$$\frac{1}{p^{(N+1)!}} < \left| \alpha - \frac{r_N}{p^{N!}} \right| < \frac{2}{p^{(N+1)!}} \quad \text{and} \quad \frac{1}{q^{(N+1)!}} < \left| \beta - \frac{s_N}{q^{N!}} \right| < \frac{2}{q^{(N+1)!}}. \qquad (1.2)$$

The polynomial $f(x, y)$ can be expressed as

$$f(x, y) = \sum_{I} C_I \, (x - \alpha)^i \, (y - \beta)^j$$

where $I = (i,j)$ runs over all pairs of non-negative integers and C_I's are real numbers which are zero for all but finitely many I's. Each such pair I gives a distinct integer $d_I := p^i q^j$. Among all the pairs I for which $C_I \neq 0$, let $I_0 = (i_0, j_0)$ be such that d_{I_0} is minimal. Then by (1.1) and (1.2), for all large N,

$$0 < |C_{I_0}| \leq \sum_{I \neq I_0} |C_I| \, 2^D \left(\frac{d_{I_0}}{d_I} \right)^{(N+1)!}$$

where D is the total degree of the polynomial $f(x,y)$. Since $d_{I_0} < d_I$, by choosing N large enough, we arrive at a contradiction. Thus $f(R_N, S_N) \neq 0$ for infinitely many N. For each such N, we have by Exercise 2

$$|(\alpha, \beta) - (R_N, S_N)| \geq \frac{c}{p^{N!D}}.$$

Here $|.|$ is the standard Euclidean norm on $\mathbb{R}^2$. On the other hand, (1.2) gives

$$|(\alpha, \beta) - (R_N, S_N)| \leq \frac{c_0}{q^{(N+1)!}}.$$

As $N \to \infty$, this contradicts the lower bound above. $\square$

Evidently, instead of taking two coprime numbers, we can generalise the above theorem by considering finitely many multiplicatively independent integers (see Theorem 2 of [1]).

Exercises

1. Show that Liouville's theorem holds for complex algebraic numbers of degree $n \geq 2$.

2. Let $f \in \mathbb{Z}[x_1, \ldots, x_n]$ be a non-zero polynomial with degree d_i in variable x_i. Suppose that $f(\overline{\alpha}) = 0$ for some $\overline{\alpha} \in \mathbb{R}^n$. Then show that there exists a constant $c > 0$ depending on $\overline{\alpha}$ and f such that for any $\overline{\beta} = (a_1/b_1, \ldots, a_n/b_n) \in \mathbb{Q}^n$, either

$$f(\overline{\beta}) = 0$$

   or

$$|\overline{\alpha} - \overline{\beta}| \geq \frac{c}{b_1^{d_1} \cdots b_n^{d_n}}.$$

   Here $|.|$ is the standard Euclidean norm on $\mathbb{R}^n$.

3. Show that the set of algebraic numbers is countable.

4. Show that $\frac{1+\sqrt{-3}}{2}$ is an algebraic integer.

5. Find the minimal polynomial of $\sqrt{2} + \sqrt{3}$ over $\mathbb{Q}$.

6. Show that the number
$$\sum_{n=0}^{\infty} \frac{1}{2^{n^2}}$$
is irrational.

7. Show that there are uncountably many Liouville numbers.

8. Show that the reciprocal of a Liouville number is again a Liouville number.

9. Show that every real number is expressible as a sum of two Liouville numbers.

Chapter 2

Hermite's Theorem

We will begin with the proof that e is transcendental, a result first proved by Charles Hermite in 1873.

Theorem 2.1 *e is transcendental.*

Proof. We make the observation that for a polynomial f and a complex number t,

$$\int_0^t e^{-u} f(u) du = [-e^{-u} f(u)]_0^t + \int_0^t e^{-u} f'(u) du$$

which is easily seen on integrating by parts. Here the integral is taken over the line joining 0 and t. If we let

$$I(t, f) := \int_0^t e^{t-u} f(u) du,$$

then we see that

$$I(t, f) = e^t f(0) - f(t) + I(t, f').$$

If f is a polynomial of degree m, then iterating this relation gives

$$I(t, f) = e^t \sum_{j=0}^m f^{(j)}(0) - \sum_{j=0}^m f^{(j)}(t). \tag{2.1}$$

If F is the polynomial obtained from f by replacing each coefficient of f by its absolute value, then it is easy to see from the definition of $I(t, f)$ that

$$|I(t, f)| \le |t| e^{|t|} F(|t|). \tag{2.2}$$

M.R. Murty and P. Rath, *Transcendental Numbers*, DOI 10.1007/978-1-4939-0832-5_2, © Springer Science+Business Media New York 2014

With these observations, we are now ready to prove the theorem. Suppose e is algebraic of degree n. Then

$$a_n e^n + a_{n-1} e^{n-1} + \cdots + a_1 e + a_0 = 0 \qquad (2.3)$$

for some integers a_i and $a_0 a_n \neq 0$. We will consider the combination

$$J := \sum_{k=0}^{n} a_k I(k, f)$$

with

$$f(x) = x^{p-1} (x-1)^p \cdots (x-n)^p$$

where $p > |a_0|$ is a large prime. Using (2.3), we see that

$$J = -\sum_{j=0}^{m} \sum_{k=0}^{n} a_k f^{(j)}(k)$$

where $m = (n+1)p - 1$. Since f has a zero of order p at $1, 2, \ldots, n$ and a zero of order $p-1$ at 0, we have that the summation actually starts from $j = p-1$. For $j = p-1$, the contribution from f is

$$f^{(p-1)}(0) = (p-1)!(-1)^{np} n!^p.$$

Thus if $n < p$, then $f^{(p-1)}(0)$ is divisible by $(p-1)!$ but not by p. If $j \geq p$, we see that $f^{(j)}(0)$ and $f^{(j)}(k)$ are divisible by $p!$ for $1 \leq k \leq n$. Hence J is a non-zero integer divisible by $(p-1)!$ and consequently

$$(p-1)! \leq |J|.$$

On the other hand, our estimate (2.2) shows that

$$|J| \leq \sum_{k=0}^{n} |a_k| e^k F(k) k \leq A n e^n (2n)!^p$$

where A is the maximum of the absolute values of the a_k's. The elementary observation

$$e^p \geq \frac{p^{p-1}}{(p-1)!}$$

gives

$$p^{p-1} e^{-p} \leq (p-1)! \leq |J| \leq A n e^n (2n)!^p.$$

For p sufficiently large, this is a contradiction. $\square$

Exercises

1. Show that for any polynomial f, we have

$$\int_0^\pi f(x) \sin x \, dx = f(\pi) + f(0) - \int_0^\pi f''(x) \sin x \, dx.$$

2. Utilise the identity in the previous exercise to show π is irrational as follows. Suppose $\pi = a/b$ with a, b coprime integers. Let

$$f(x) = \frac{x^n(a - bx)^n}{n!}.$$

Prove that

$$\int_0^\pi f(x) \sin x \, dx$$

is a non-zero integer and derive a contradiction from this.

3. Use Euler's identity

$$\sum_{n=1}^\infty \frac{1}{n^2} = \frac{\pi^2}{6}$$

to prove that there are infinitely many primes.

4. Use the series $\sum_{n=0}^\infty 1/n!$ to show that e is irrational.

5. Show that e is not algebraic of degree 2 by considering the relation

$$Ae + Be^{-1} + C = 0, \quad A, B, C \in \mathbb{Z},$$

and using the infinite series for e and e^{-1} and arguing as in the previous exercise.

6. Prove that $e^{\sqrt 2}$ is irrational (Hint: Consider the series expansion for $\alpha = e^{\sqrt 2} + e^{-\sqrt 2}$).

Chapter 3

Lindemann's Theorem

We will now prove that π is transcendental. This was first proved by
F. Lindemann in 1882 by modifying Hermite's methods. The proof proceeds
by contradiction. Before we begin the proof, we recall two facts from algebraic
number theory. The first is that if α is an algebraic number with minimal
polynomial over $\mathbb{Z}$ given by

$$a_n x^n + a_{n-1} x^{n-1} + \cdots + a_1 x + a_0,$$

then $a_n \alpha$ is an algebraic integer. Indeed, if we multiply the polynomial by a_n^{n-1},
we see that

$$(a_n \alpha)^n + a_{n-1}(a_n \alpha)^{n-1} + \cdots + a_0 a_n^{n-1} = 0$$

thereby showing that $a_n \alpha$ satisfies a monic polynomial equation with integer
coefficients. The other roots of the minimal polynomial of α are called the
conjugates of α. Sometimes we write these conjugates as

$$\alpha^{(1)}, \alpha^{(2)}, \ldots, \alpha^{(n)}$$

with $\alpha^{(1)} = \alpha$. The second fact which we require is from Galois theory and the
symmetric polynomial theorem. More precisely, let $f(x_1, \ldots, x_n)$ be a symmetric polynomial in $\mathbb{Q}[x_1, \ldots, x_n]$, that is,

$$f(x_1, \ldots, x_n) = f(x_{\sigma(1)}, \ldots, x_{\sigma(n)})$$

for any element σ of the symmetric group S_n. If α is an algebraic number of
degree n with conjugates $\alpha = \alpha_1, \ldots, \alpha_n$, then $f(\alpha_1, \ldots, \alpha_n) \in \mathbb{Q}$. Furthermore,
if α is an algebraic integer and f has integer coefficients, then $f(\alpha_1, \ldots, \alpha_n)$ is
necessarily an integer (see Exercise 1). With these remarks, we can proceed to
the proof that π is transcendental.

M.R. Murty and P. Rath, *Transcendental Numbers*, DOI 10.1007/978-1-4939-0832-5_3,   11
© Springer Science+Business Media New York 2014

Theorem 3.1 (Lindemann, 1882 [80]) π *is transcendental.*

Proof. Suppose not. Then $\alpha = i\pi$ is also algebraic. Let α have degree d and let $\alpha = \alpha_1, \ldots, \alpha_d$ be the conjugates. Let N be the leading coefficient of the minimal polynomial of α over $\mathbb{Z}$. By our remarks before, $N\alpha$ is an algebraic integer. Since

$$e^{i\pi} = -1,$$

we have

$$(1 + e^{\alpha_1})(1 + e^{\alpha_2}) \cdots (1 + e^{\alpha_d}) = 0.$$

The product can be written out as a sum of 2^d terms of the form e^{θ} where

$$\theta = \epsilon_1 \alpha_1 + \cdots + \epsilon_d \alpha_d, \quad \epsilon_i = 0, 1.$$

Suppose that exactly n of these numbers are non-zero and denote them by $\beta_1, \ldots, \beta_n$. Note that these numbers constitute all the roots of a polynomial with integer coefficients. To see this, it suffices to observe that the polynomial

$$\prod_{\epsilon_1 = 0}^{1} \cdots \prod_{\epsilon_d = 0}^{1} (x - (\epsilon_1 \alpha_1 + \cdots + \epsilon_d \alpha_d))$$

is symmetric in $\alpha_1, \ldots, \alpha_d$ and hence lies in $\mathbb{Q}[x]$. The roots of this polynomial are $\beta_1, \ldots, \beta_n$ and 0 which has multiplicity $a = 2^d - n$. Dividing by x^a and clearing the denominator, we get a polynomial in $\mathbb{Z}[x]$ with roots $\beta_1, \ldots, \beta_n$. Now

$$(1 + e^{\alpha_1})(1 + e^{\alpha_2}) \cdots (1 + e^{\alpha_d}) = 0$$

which implies

$$(2^d - n) + e^{\beta_1} + \cdots + e^{\beta_n} = 0.$$

With $I(t, f)$ as in the previous chapter, we consider the combination

$$K := I(\beta_1, f) + \cdots + I(\beta_n, f)$$

where

$$f(x) = N^{np} x^{p-1} (x - \beta_1)^p \cdots (x - \beta_n)^p$$

and p denotes again a large prime. Thus,

$$K = -(2^d - n) \sum_{j=0}^{m} f^{(j)}(0) - \sum_{j=0}^{m} \sum_{k=1}^{n} f^{(j)}(\beta_k)$$

where $m = (n + 1)p - 1$. The sum over k is a symmetric function in $N\beta_1, \ldots, N\beta_n$. Noting that $N\beta_1, \ldots, N\beta_n$ are all the roots of a monic polynomial over the integers, we conclude that the summation is a rational integer. Moreover, the derivatives $f^{(j)}(\beta_k)$ vanish for $j < p$ and the summation for fixed $j \geq p$ is divisible by $p!$. Also for p sufficiently large,

$$f^{(p-1)}(0) = (p - 1)! (-N)^{np} (\beta_1 \cdots \beta_n)^p$$

is not divisible by p. In addition, $f^{(j)}(0)$ is divisible by $p!$ for $j \geq p$. As in the previous chapter, let F be the polynomial obtained from f by replacing each coefficient of f by its absolute value. Proceeding as before, we find that

$$|K| \leq \sum_{k=1}^{n} |\beta_k| e^{|\beta_k|} F(|\beta_k|) \leq AC^p$$

for some constants A and C. On the other hand, K is a non-zero rational integer divisible by $(p-1)!$ and hence must be at least as large in absolute value. Comparing the growth as p tends to infinity gives us the desired contradiction. $\square$

Exercises

1. Let α be an algebraic integer of degree n with conjugates $\alpha = \alpha_1, \ldots, \alpha_n$. Suppose for $k \geq 0$,

$$F = F(x_1, \ldots, x_k; \alpha_1, \ldots, \alpha_n) \in \mathbb{Z}[x_1, \ldots, x_k; \alpha_1, \ldots, \alpha_n].$$

   Further suppose that F is a symmetric function in $\alpha_1, \ldots, \alpha_n$ with coefficients in the ring $\mathbb{Z}[x_1, \ldots, x_k]$. Then show that $F \in \mathbb{Z}[x_1, \ldots, x_k]$ when $k > 0$ and $F \in \mathbb{Z}$ when $k = 0$.

2. A real number α is said to be *constructible* if, by means of a straightedge, a compass, and a line segment of length 1, we can construct a line segment of length $|\alpha|$ in a finite number of steps. Show that if α, β are constructible, then so are $\alpha+\beta$, $\alpha-\beta$, $\alpha\beta$ and α/β for $\beta \neq 0$. Thus the set of constructible numbers forms a subfield of the reals.

3. Show that if α is constructible, so is $\sqrt{|\alpha|}$. [Hint: consider the circle of diameter $|\alpha| + 1$ with center $(\frac{1}{2}(1 + |\alpha|), 0)$ in $\mathbb{R}^2$ and consider the intersection of the perpendicular drawn at $(1, 0)$ and the circle.]

4. Let F be any subfield of the reals. Call $F \times F$ the *plane* of F and any line joining two points in the plane of F a *line in F*. A circle whose center is in the plane of F and whose radius is in F will be called a *circle in F*. Show that a line F is defined by the equation

$$ax + by + c = 0, \quad a, b, c \in F$$

   and a circle in F is defined by the equation

$$x^2 + y^2 + ax + by + c = 0, \quad a, b, c \in F.$$

5. From the previous exercise deduce that any constructible number must necessarily be an algebraic number. Deduce using Lindemann's theorem that $\sqrt{\pi}$ is not constructible. Hence, it is impossible to construct using a straightedge and compass a square whose area is equal to π.

6. Show that $\cos(\pi/9)$ is algebraic of degree 3 over the rationals.

7. Using Exercise 4, show that a constructible number must necessarily have degree a power of 2 over the rationals. Conclude that $\pi/3$ cannot be trisected using straightedge and compass.

8. Show that if α is constructible, then it lies in a subfield of $\mathbb{R}$ obtained from $\mathbb{Q}$ by a finite sequence of quadratic extensions.

9. Let p be a prime number. Show that if a regular p-gon can be constructed using straightedge and compass, then $p - 1 = 2^r$ for some r. [The converse is also true but more difficult to prove and is a celebrated "teenage" theorem of Gauss.]

Chapter 4

The Lindemann–Weierstrass Theorem

In 1882, Lindemann wrote a paper in which he sketched a general result, special cases of which imply the transcendence of e and π. This general result was later proved rigorously by K. Weierstrass in 1885. Before we begin, we make some remarks pertaining to algebraic number theory. Let K be an algebraic number field. The collection of algebraic integers in K forms a ring, denoted $\mathcal{O}_K$, and is called the ring of integers of K. The theorem of the primitive element shows there exists a θ such that $K = \mathbb{Q}(\theta)$. If $\theta^{(1)}, \ldots, \theta^{(r)}$ are all the conjugates of θ, then one speaks of the *conjugate fields* $K^{(i)} := \mathbb{Q}(\theta^{(i)})$. This gives rise to an isomorphism σ_i of fields $K \simeq K^{(i)}$ given by the map $\sigma_i(\theta) = \theta^{(i)}$, which is then extended to all of K in the obvious way.

Theorem 4.1 (Lindemann–Weierstrass, 1885) *If $\alpha_1, \ldots, \alpha_s$ are distinct algebraic numbers, then $e^{\alpha_1}, \ldots, e^{\alpha_s}$ are linearly independent over $\overline{\mathbb{Q}}$.*

Proof. Suppose that we have

$$d_1 e^{\alpha_1} + \cdots + d_s e^{\alpha_s} = 0 \qquad (4.1)$$

for some algebraic numbers $d_1, \ldots, d_s$ not all zero. By multiplying an appropriate rational integer, we may assume that $d_1, \ldots, d_s$ are algebraic integers. Further, multiplying the above equation with equations of the form $\sum_{j=1}^{s} \sigma_k(d_j) e^{\alpha_j}$ for all the embeddings σ_k of the field $\mathbb{Q}(d_1, \ldots, d_s)$, we may assume a relation of the form

$$a_1 e^{\gamma_1} + \cdots + a_n e^{\gamma_n} = 0 \qquad (4.2)$$

M.R. Murty and P. Rath, *Transcendental Numbers*, DOI 10.1007/978-1-4939-0832-5_4,
© Springer Science+Business Media New York 2014

where a_i's are rational integers and γ_i's are distinct algebraic numbers. We may further assume that each conjugate of γ_i is also included in the above list of algebraic integers. Now if we let K to be the algebraic number field generated by $\gamma_1, \ldots, \gamma_n$ and all their conjugates, then it is natural to consider the "conjugate" functions for real variable t,

$$A_i(t) := a_1 e^{\gamma_1^{(i)} t} + \cdots + a_n e^{\gamma_n^{(i)} t}.$$

Since the γ_j are distinct, these functions are not identically zero (see Exercise 1). If we let

$$B(t) = \prod_i A_i(t) = b_1 e^{\beta_1 t} + \cdots + b_M e^{\beta_M t},$$

where the product is over all the conjugate functions, then it is clear that the Taylor coefficients of $B(t)$ are symmetric functions in all of the conjugates and so are rational numbers by our earlier remarks. Moreover, the b_i are rational integers not all equal to zero. Let N be an integer so that $N\beta_1, \ldots, N\beta_M$ are algebraic integers. We now proceed as in the earlier chapters. Consider the combination

$$J_r := \sum_{k=1}^{M} b_k I(\beta_k, f_r)$$

where

$$f_r(x) = N^{Mp} \frac{(x - \beta_1)^p (x - \beta_2)^p \cdots (x - \beta_M)^p}{(x - \beta_r)}$$

for $1 \le r \le M$. It is clear that $f(x) = f_1(x) + \cdots + f_M(x)$ is invariant under Galois action and hence has rational integer coefficients. Now using (2.1), we see that since $B(1) = 0$, we have

$$J_r = -\sum_{k=1}^{M} b_k \sum_{j=0}^{m} f_r^{(j)}(\beta_k),$$

where m is the degree of f_r. Arguing as in the earlier chapters, we note that the product $J_1 \cdots J_M$ is a Galois invariant algebraic integer, hence an integer. Further, it is divisible by $(p-1)!$, but not by p for suitably chosen large p. In the other direction, since each $|J_r| \le (c_r)^p$ for suitable c_r, we have

$$(p-1)! \le C^p$$

for some constant C. This gives a contradiction for large enough p which completes the proof. $\square$

The Lindemann–Weierstrass theorem generalises both the Hermite and Lindemann's theorems. Indeed, choosing $\alpha_1 = 0$ and $\alpha_2 = 1$, we retrieve Hermite's theorem that e is transcendental. Choosing $\alpha_1 = 0$ and $\alpha_2 = i\pi$, we deduce Lindemann's theorem. We also have the following corollaries.

Corollary 4.2 *If $\alpha \neq 0$ is algebraic, then e^α is transcendental.*

Proof. Take $\alpha_1 = 0$ and $\alpha_2 = \alpha$ in Theorem 4.1. $\square$

Corollary 4.3 *If $\alpha \neq 0, 1$ is algebraic, then $\log \alpha$ is transcendental.*

Proof. This is immediate from the previous corollary. $\square$

Recall that a collection of n complex numbers $\beta_1, \ldots, \beta_n$ is *algebraically independent* if there is no non-zero polynomial $P(x_1, \ldots, x_n) \in \mathbb{Z}[x_1, \ldots, x_n]$ such that

$$P(\beta_1, \ldots, \beta_n) = 0.$$

We can deduce from Theorem 4.1 the following assertion.

Theorem 4.4 *If $\alpha_1, \ldots, \alpha_n$ are algebraic numbers that are linearly independent over $\mathbb{Q}$, then*

$$e^{\alpha_1}, \ldots, e^{\alpha_n}$$

are algebraically independent.

Proof. Suppose that

$$e^{\alpha_1}, \ldots, e^{\alpha_n}$$

are algebraically dependent. Then we have

$$\sum_{i_1, \ldots, i_n} a_{i_1, \ldots, i_n} e^{i_1 \alpha_1 + \cdots + i_n \alpha_n} = 0,$$

for certain integers $a_{i_1, \ldots, i_n}$ with not all $a_{i_1, \ldots, i_n}$ equal to zero. By Theorem 4.1, the numbers

$$i_1 \alpha_1 + \cdots + i_n \alpha_n$$

cannot all be distinct. Thus $\alpha_1, \ldots, \alpha_n$ are linearly dependent over $\mathbb{Q}$. $\square$

S. Schanuel has conjectured that if $\alpha_1, \ldots, \alpha_n$ are complex numbers that are linearly independent over $\mathbb{Q}$, then the transcendence degree of the field

$$\mathbb{Q}(\alpha_1, \ldots, \alpha_n, \ e^{\alpha_1}, \ldots, e^{\alpha_n})$$

over $\mathbb{Q}$ is at least n. One consequence of this conjecture is that e and π are algebraically independent. To see this, consider the field generated by $1, 2\pi i, e, e^{2\pi i}$ over the rationals. Schanuel's conjecture predicts that the transcendence degree of this field is at least 2, which means that e and π are algebraically independent.

Schanuel's conjecture is one of the central conjectures in the theory of transcendental numbers. We will discuss this conjecture and its many implications in a later chapter.

Exercises

1. If $\alpha_1, \ldots, \alpha_n$ are distinct complex numbers, show that the function

$$a_1 e^{\alpha_1 t} + \cdots + a_n e^{\alpha_n t}$$

   is not identically zero whenever the a_i's are not all zero.

2. Show that for any non-zero algebraic number, $\sin \alpha$, $\cos \alpha$, $\tan \alpha$ are transcendental numbers. Show the same is true for the hyperbolic functions, $\sinh \alpha$, $\cosh \alpha$ and $\tanh \alpha$.

3. Show that $\arcsin \alpha$ is transcendental for any non-zero algebraic number α.

4. If $c_0, c_1, \ldots$ is a periodic sequence of algebraic numbers not all zero, then show that the series

$$\sum_{n=0}^{\infty} c_n \frac{z^n}{n!}$$

   is transcendental for any non-zero algebraic value of z.

5. Show that at least one of $\pi + e$, πe is transcendental.

Chapter 5

The Maximum Modulus Principle and Its Applications

The maximum modulus principle constitutes an essential tool in transcendence theory. Let us begin with a proof of this fundamental result. We fix the convention that a function is analytic in a closed set C if it is analytic in an open set containing C. A region is an open connected set. We consider the following version of the maximum modulus principle. The statement is not the most general, but suffices for our applications.

Theorem 5.1 (The Maximum Modulus Principle) *If f is a non-constant analytic function in a region R, then the function $|f|$ does not attain its maximum in R. In other words if for some $z_0 \in R$, $|f(z)| \leq |f(z_0)|$ for all points $z \in R$, then f is constant.*

Proof. We give two proofs of the theorem. For the first proof we use the fact that a non-constant analytic map in a region is an open map. Let $|f(z_0)| = M$. Since $|f(z)| \leq |f(z_0)|$ for all points $z \in R$, the image set $f(R)$ is contained in the closed disc $\{z : |z| \leq M\}$ and intersects the boundary. Hence $f(R)$ is not open, a contradiction.

For the second proof, for the point z_0 in R, consider the Taylor expansion of f about z_0:

$$f(z_0 + re^{i\theta}) = \sum_{n=0}^{\infty} a_n r^n e^{in\theta}.$$

M.R. Murty and P. Rath, *Transcendental Numbers*, DOI 10.1007/978-1-4939-0832-5_5,   19
© Springer Science+Business Media New York 2014

Parseval's formula (or just by noting that term by term integration is allowed as the series converges normally for a fixed r) yields that

$$\frac{1}{2\pi} \int_0^{2\pi} \left|f(z_0 + re^{i\theta})\right|^2 d\theta = \sum_{n=0}^{\infty} |a_n|^2 r^{2n}.$$

Thus if z_0 is a point where the maximum is attained, we have $|a_0| = M$ and

$$M^2 = |a_0|^2 \leq |a_0|^2 + |a_1|^2 r^2 + \cdots \leq |f(z_0)|^2 = M^2$$

so that we are forced to have $a_1 = a_2 = \cdots = 0$ and f is constant. $\square$

We shall apply the maximum modulus principle mostly to the following special case: if f is continuous in the closed disc $\{z : |z| \leq R\}$ and analytic in the interior, then the maximum of $|f|$ in the closed disc is necessarily attained on the boundary. The principle can be used to prove the fundamental theorem of algebra.

Corollary 5.2 (The Fundamental Theorem of Algebra) *If $n \geq 1$ and*

$$f(z) = a_n z^n + a_{n-1} z^{n-1} + \cdots + a_1 z + a_0$$

is a polynomial with complex coefficients and $a_n \neq 0$, then f has precisely n roots over the complex numbers.

Proof. It suffices to show that f has at least one root for then, we can apply the division algorithm to reduce the degree of f and apply induction. If $f(z) \neq 0$ for every complex value of z, then $1/f$ is entire. We will apply the maximum modulus principle to $1/f$. Clearly $1/f(z)$ tends to zero as $|z|$ tends to infinity. Thus for any given α, there exists an R such that

$$\frac{1}{|f(z)|} < \frac{1}{|f(\alpha)|}$$

for $|z| \geq R$. But we can choose R sufficiently large so as to ensure that $|\alpha| < R$. This violates the maximum modulus principle applied to the non-constant function $1/f$. Thus f has a root in $\mathbb{C}$. $\square$

The following two corollaries suggest that the inequality in the maximum modulus principle can be improved if we have knowledge of zeros of the function lying inside a disc of radius R.

Corollary 5.3 (Schwarz's Lemma) *Suppose that f is analytic in the closed disc $\{z : |z| \leq R\}$ and $f(0) = 0$. Then in this disc,*

$$|f(z)| \leq |f|_R (|z|/R),$$

where $|f|_R$ is maximum of $|f|$ on the circle $\{z : |z| = R\}$ of radius R.

Proof. The function $g(z) = f(z)/z$ initially defined for $0 < |z| \leq R$ can be analytically extended to $|z| \leq R$. Applying the maximum modulus principle to g gives the result immediately. $\square$

Corollary 5.4 (Jensen's Inequality) *Let f be analytic in $\{z : |z| \leq R\}$ and $f(0) \neq 0$. If the zeros of f in the open disc are $z_1, z_2, \ldots, z_N$, with each zero being repeated according to multiplicity, then*

$$|f(0)| \leq |f|_R (|z_1 \cdots z_N|/R^N).$$

Proof. It is easily seen (see Exercise 1 below) that

$$\frac{R^2 - z\overline{z_n}}{R(z - z_n)}$$

has absolute value 1 for $|z| = R$. Thus, the function

$$g(z) = f(z) \prod_{n=1}^{N} \frac{R^2 - z\overline{z_n}}{R(z - z_n)}$$

is analytic on the closed disc of radius R and

$$|g(z)| = |f(z)|$$

for $|z| = R$. The maximum modulus principle implies

$$|g(z)| \leq |f|_R.$$

Putting $z = 0$ gives the result. $\square$

Corollary 5.5 *Let f be as in the previous corollary. Let for positive r, $\nu(r) = \nu(f, r)$ denote the number of zeros of f in the open disc $\{z : |z| < r\}$ counted according to multiplicity. Then,*

$$\int_0^R \frac{\nu(x)}{x} dx \leq \log |f|_R - \log |f(0)|.$$

Proof. Since

$$\log \frac{R^N}{|z_1 \cdots z_N|} = \sum_{n=1}^{N} \int_{|z_n|}^{R} \frac{dx}{x} = \int_0^R \frac{\nu(x)}{x} dx,$$

the result is immediate from the previous corollary. $\square$

One of the main consequences of these results is a relationship between the number of zeros in a disc and the rate of growth of the function. We say that an entire function f is of *strict order* $\leq \rho$ for a positive real ρ if there is a constant $C > 0$ such that for any positive R,

$$|f(z)| \leq C^{R^\rho} \quad \text{whenever } |z| \leq R.$$

If f is as above, then the greatest lower bound of all ρ for which the above condition holds is called the *order* of f.

Corollary 5.6 *If f is a non-zero analytic function of strict order $\leq \rho$, then the number of zeros of f inside the disc of radius $R \geq 1$ is bounded by AR^ρ where A is a constant that depends on f and not on R.*

Proof. Suppose that f has a zero of order n at $z = 0$. Then, $g(z) = f(z)/z^n$ is analytic at $z = 0$ and $g(0) \neq 0$. Applying the previous corollary, we see that

$$\nu(g, R) \log 2 \leq \log |g|_{2R} - \log |g(0)|.$$

Thus,

$$\nu(g, R) \log 2 + n \log R \leq \log |f|_{2R} - \log |g(0)|$$

from which the result easily follows. $\square$

Exercises

1. Show that for $|w| < R$, the quotient

$$\frac{R^2 - z\overline{w}}{R(z - w)}$$

has absolute value 1 for $|z| = R$.

2. Let f be a function analytic in $|z| \leq R$ and non-vanishing there. Show that the minimum modulus of f, $\min_{|z| \leq R} |f(z)|$ is attained on the boundary.

3. Let f be analytic in $|z| \leq R$ with $R > 0$. Show that

$$|f^{(n)}(0)| \leq n!|f|_R/R^n.$$

4. Deduce from the previous exercise that a bounded entire function is constant. (This is a famous theorem of Liouville.)

5. Deduce the fundamental theorem of algebra from Liouville's theorem.

6. Suppose f is analytic in $|z| < 1$ with $|f(z)| \leq 1$ and $f(0) = 0$. Then show that $|f'(0)| \leq 1$ with equality if and only if $f(z) = cz$ where $|c| = 1$.

7. If f and g are entire functions of order ρ_1 and ρ_2 respectively, show that the function fg is of order ρ with $\rho \leq \max(\rho_1, \rho_2)$. Further, if $\rho_1 \neq \rho_2$, then show that fg has order equal to $\max(\rho_1, \rho_2)$.

8. Let f and g be as in the previous exercise. What can you conclude about the order of $f + g$?

Chapter 6

Siegel's Lemma

The following lemma is a fundamental tool in transcendental number theory.

Lemma 6.1 (Siegel) *Let a_{ij} be integers of absolute value at most A for $1 \leq i \leq r, 1 \leq j \leq n$. Consider the homogeneous system of r equations*

$$\sum_{j=1}^{n} a_{ij} x_j = 0, \quad 1 \leq i \leq r$$

in n unknowns. If $n > r$, there is a non-trivial integral solution satisfying

$$|x_j| \leq B$$

where

$$B = 2(2nA)^{\frac{r}{n-r}}.$$

Proof. Let $C = (a_{ij})$ be the matrix associated with the system of equations. Then C maps $\mathbb{R}^n$ into $\mathbb{R}^r$. Moreover, it maps $\mathbb{Z}^n$ into $\mathbb{Z}^r$. Let $H \geq 1$ be a real number and $\mathbb{Z}^n(H)$ be the set of vectors in $\mathbb{R}^n$ with integral co-ordinates of absolute value at most H. Then clearly C maps $\mathbb{Z}^n(H)$ into $\mathbb{Z}^r(nAH)$. If

$$(2nAH + 1)^r < (2H)^n,$$

then the map cannot be injective. In particular if

$$(2H)^{n/r} \geq (2H)(2nA) > 2nAH + 1,$$

M.R. Murty and P. Rath, *Transcendental Numbers*, DOI 10.1007/978-1-4939-0832-5_6,   23
© Springer Science+Business Media New York 2014

then there will be at least two distinct vectors mapping to the same point. The difference of these two vectors gives a solution to the homogeneous system satisfying

$$|x_j| \leq 2H.$$

Choosing $H = (2nA)^{\frac{r}{n-r}}$ gives the result. $\square$

We will need a generalisation of this lemma to number fields. To this end, we review some basic algebraic number theory. Let K be a number field. If $\alpha \in K$ is an algebraic number, then consider the set of all integers m such that $m\alpha$ is an algebraic integer. The set of such integers contains a non-zero integer by our earlier remarks (see Chap. 3). Moreover, it is an ideal of $\mathbb{Z}$ and hence principal. The positive generator of this ideal will be called the *denominator* of α and denoted $d(\alpha)$. We say d is the *denominator* of the algebraic numbers $\alpha_1, \ldots, \alpha_n$ if d is the least common multiple of the numbers $d(\alpha_1), \ldots, d(\alpha_n)$. We will also define the *height* of α, denoted $H(\alpha)$, to be the maximum absolute value of all its conjugates.

The second fact we need is that the ring of integers $\mathcal{O}_K$ of a number field K of degree t has an *integral basis*. That is, there are algebraic integers $\omega_1, \ldots, \omega_t \in \mathcal{O}_K$ such that every element of $\mathcal{O}_K$ can be written as

$$a_1\omega_1 + \cdots + a_t\omega_t$$

with $a_i \in \mathbb{Z}$. Let $\sigma_1, \ldots, \sigma_t$ be the embeddings of K in $\mathbb{C}$ and for any $\omega \in K$, $\omega^{(j)} = \sigma_j(\omega)$ denote its j-th conjugate. The $t \times t$ matrix whose (i,j)-th entry is $\omega_i^{(j)}$ is easily verified to be invertible.

Lemma 6.2 *Let $\alpha_{ij} \in \mathcal{O}_K$ be algebraic integers of height at most A for $1 \leq i \leq r, 1 \leq j \leq n$. Consider the homogeneous system of r equations*

$$\sum_{j=1}^{n} \alpha_{ij} x_j = 0, \quad 1 \leq i \leq r$$

in n unknowns. If $n > r$, there is a non-trivial $\mathcal{O}_K$-integral solution satisfying

$$H(x_j) \leq B$$

where

$$B = C(CnA)^{\frac{r}{n-r}}.$$

Here C is an absolute constant that depends only on K.

Proof. Let t be the degree of the number field K. We write each of the numbers α_{ij} in terms of an integral basis:

$$\alpha_{ij} = \sum_{k=1}^{t} a_{ijk} \omega_k, \quad a_{ijk} \in \mathbb{Z}.$$

From these equations, we also see that by inverting the $t \times t$ matrix

$$(\omega_k^{(\ell)}),$$

we can solve for the a_{ijk}. Thus we see that

$$|a_{ijk}| \leq C_0 A$$

where C_0 is a constant depending only on K (more precisely, on the integral basis ω_i's and the degree t). We write each x_j as

$$x_j = \sum_\ell y_{j\ell} \omega_\ell$$

so that the system becomes

$$\sum_{j=1}^{n} \sum_{k=1}^{t} \sum_{\ell=1}^{t} a_{ijk} y_{j\ell} \omega_k \omega_\ell = 0$$

with $y_{j\ell}$ to be solved in $\mathbb{Z}$. We can write

$$\omega_k \omega_\ell = \sum_{m=1}^{t} c_{k\ell m} \omega_m, \quad c_{k\ell m} \in \mathbb{Z}.$$

Thus we have,

$$\sum_{m=1}^{t} \sum_{j=1}^{n} \sum_{k,\ell=1}^{t} a_{ijk} y_{j\ell} c_{k\ell m} \omega_m = 0.$$

Since the ω_m's are linearly independent over $\mathbb{Q}$, the original system is now equivalent to a new system of equations with ordinary integer coefficients in the unknowns $y_{j\ell}$. More precisely, we get the following homogeneous system of rt equations

$$\sum_{j=1}^{n} \sum_{\ell=1}^{t} \sum_{k=1}^{t} a_{ijk} c_{k\ell m} y_{j\ell} = 0, \quad 1 \leq i \leq r, \; 1 \leq m \leq t$$

in the nt unknowns $y_{j\ell}$. We can now apply the previous lemma and obtain the desired result by suitably choosing C which depends only on K. $\square$

We will need one more variation of the previous lemma that will allow the coefficients to be algebraic numbers instead of algebraic integers.

Lemma 6.3 *Let $\alpha_{ij} \in K$ be algebraic numbers of height at most A for $1 \leq i \leq r$, $1 \leq j \leq n$. Consider the homogeneous system of r equations*

$$\sum_{j=1}^{n} \alpha_{ij} x_j = 0, \quad 1 \leq i \leq r$$

in n unknowns. Let d_i be the denominator for the coefficients of the i-th equation and let d be the maximum of the d_i's. If $n > r$, there exists a non-trivial $\mathcal{O}_K$-integral solution satisfying

$$H(x_j) \leq B$$

where

$$B = C(CndA)^{\frac{r}{n-r}}$$

with C an absolute constant depending only on K.

Proof. We simply multiply the i-th equation by d_i and then apply the previous lemma. $\square$

Exercises

1. Let K be an algebraic number field and $\omega_1, \ldots, \omega_n$ an integral basis for the ring of integers $\mathcal{O}_K$. Show that the matrix $(\omega_j^{(i)})$ is invertible.

2. Let α be an algebraic integer such that the height of α is equal to one. Show that α is a root of unity.

3. A real algebraic integer $\alpha > 1$ is called a Pisot–Vijayaraghavan number (or a PV number) if all its other conjugates have absolute value strictly less than one. Show that there are infinitely many PV numbers.
   (Hint: Try using Rouché's theorem in complex analysis.)

4. For any real x, let $||x||$ denote its distance from the nearest integer. Show that if α is a PV number, then the sequence $||\alpha^n||$ tends to zero as n tends to infinity. (It is a longstanding conjecture that the converse is also true, that is, $||\alpha^n|| \to 0$ for any real $\alpha > 1$ implies that α is a PV number. This is known to be true if α is assumed to be algebraic.)

5. Prove the following sharpening of Siegel's lemma: let $a_{ij} \in \mathbb{Z}$ be integers satisfying

$$\sum_{j=1}^{n} |a_{ij}| \leq A_i, \quad 1 \leq i \leq r.$$

Consider the homogeneous system of r equations

$$\sum_{j=1}^{n} a_{ij} x_j = 0, \quad 1 \leq i \leq r$$

in n unknowns. If $n > r$, there is a non-trivial integer solution satisfying

$$|x_i| \leq B,$$

where

$$B = (A_1 \cdots A_r)^{1/(n-r)}.$$

Chapter 7

The Six Exponentials Theorem

In this chapter and subsequent chapters, we will use Siegel's lemma and the maximum modulus principle to prove transcendence results. We shall begin with the six exponentials theorem. The proof of this theorem involves the notion of norm of an algebraic number which we recall. Let K be a number field and Σ be the set of embeddings of K into $\mathbb{C}$. Then for any $\alpha \in K$, we define the *relative norm* $N_{K/\mathbb{Q}}(\alpha)$ of α to be

$$N_{K/\mathbb{Q}}(\alpha) = \prod_{\sigma \in \Sigma} \sigma(\alpha).$$

We refer $N(\alpha) = N_{\mathbb{Q}(\alpha)/\mathbb{Q}}(\alpha)$ *to be the* norm of α. It is clear that

$$N_{K/\mathbb{Q}}(\alpha) = N(\alpha)^d$$

where $d = [K : \mathbb{Q}(\alpha)]$. When α is an algebraic integer, its norm is a rational integer. Furthermore when $\alpha \neq 0$, we have the obvious but important inequality:

$$1 \leq |N_{K/\mathbb{Q}}(\alpha)| \leq H(\alpha)^{n-1}|\alpha|$$

where $n = [K : \mathbb{Q}]$ and $H(\alpha)$ is the height of α.

Theorem 7.1 *Let* x_1, x_2 *be two complex numbers linearly independent over* $\mathbb{Q}$. *Let* y_1, y_2, y_3 *be three complex numbers linearly independent over* $\mathbb{Q}$. *Then at least one of the six numbers*

$$\exp(x_i y_j), \quad 1 \leq i \leq 2, \quad 1 \leq j \leq 3,$$

is transcendental.

M.R. Murty and P. Rath, *Transcendental Numbers*, DOI 10.1007/978-1-4939-0832-5_7,
© Springer Science+Business Media New York 2014

Remark. There is the four exponentials conjecture of Schneider that says the theorem should still be valid if y_1, y_2, y_3 are replaced by y_1, y_2 linearly independent over $\mathbb{Q}$. We refer to the interested reader a paper of Diaz [44] where he investigates the interrelation between values of the modular j-function (which we shall be defining later) and the four-exponential conjecture.

Proof. Suppose that the conclusion of the theorem is false. Let K be an algebraic number field containing the numbers

$$\exp(x_i y_j), \quad 1 \le i \le 2, \quad 1 \le j \le 3.$$

Let d be the common denominator for these numbers. We will consider the function

$$F(z) = \sum_{i,j=1}^{r} a_{ij} e^{(ix_1 + jx_2)z}$$

where the $a_{ij} \in \mathcal{O}_K$ will be suitably chosen so that F has a zero at the points

$$k_1 y_1 + k_2 y_2 + k_3 y_3$$

with $1 \le k_i \le n$ and where n is a parameter also to be suitably chosen. This amounts to solving n^3 equations in r^2 unknowns. To apply Siegel's lemma, we need $r^2 > n^3$. The coefficients of the equations are the algebraic numbers

$$\exp((ix_1 + jx_2)(k_1 y_1 + k_2 y_2 + k_3 y_3))$$

with denominators bounded by d^{6rn}. By Siegel's lemma, we can find algebraic integers a_{ij} of height at most

$$C(Cr^2 d^{6rn} e^{corn})^{\frac{n^3}{r^2 - n^3}}.$$

We will choose $r^2 = (4n)^3$ so that the a_{ij}'s have height at most $e^{c_1 n^{5/2}}$. Since x_1, x_2 are linearly independent over $\mathbb{Q}$, we see that F is not identically zero. Moreover, F takes values in K for all integral linear combinations of y_1, y_2, y_3. Since F is of strict order ≤ 1, not all such integral linear combinations can give rise to zeros of F since the number of such zeros in a circle of radius R grows like R^3 whereas the number of possible zeros of F grows like R by Jensen's theorem discussed in Chap. 5. Alternatively, since the set of numbers $k_1 y_1 + k_2 y_2 + k_3 y_3$ is not discrete, not all of these can be zeros of F. Let s be the largest positive integer such that

$$F(k_1 y_1 + k_2 y_2 + k_3 y_3) = 0 \quad \text{for } 1 \le k_i \le s.$$

Then by construction, $s \ge n$. Let

$$w = k_1 y_1 + k_2 y_2 + k_3 y_3$$

be such that $F(w) \ne 0$ with some $k_i = s + 1$ and $1 \le k_i \le s + 1$ for all i. Let us observe that

$$d^{6r(s+1)} F(w)$$

is therefore a non-zero algebraic integer. Hence the absolute value of its norm is at least one. In addition, we have the height estimate

$$\log H(F(w)) \ll n^{5/2} + (s+1)r \ll s^{5/2}.$$

This means

$$|F(w)| \geq C^{-s^{5/2}}$$

for some positive constant C. We now show that this is a contradiction. Clearly

$$F(w) = \lim_{z \to w} F(z) \prod_{1 \leq k_1, k_2, k_3 \leq s} \left(\frac{w - (k_1 y_1 + k_2 y_2 + k_3 y_3)}{z - (k_1 y_1 + k_2 y_2 + k_3 y_3)} \right).$$

There are s^3 terms in the product and the function on the right-hand side is entire. We want to estimate the size of $F(w)$. We can apply the maximum modulus principle on the circle of radius R to the entire function on the right-hand side. We will choose R so as to ensure that $|w| < R$ and

$$|z - (k_1 y_1 + k_2 y_2 + k_3 y_3)| \geq R/2$$

for all z on the circle. Thus

$$|F(w)| \leq |F|_R (C_1 s/R)^{s^3}$$

for some constant $C_1 > 0$. But an easy estimation gives

$$|F|_R \ll e^{c_1 n^{5/2} + c_2 rR} r^2,$$

for some positive constants c_1, c_2. Putting everything together gives

$$\log |F(w)| \ll n^{5/2} + rR + s^3 \log(s/R).$$

We will choose $R = s^{3/2}$. This contradicts our earlier estimate that

$$\log |F(w)| \gg -s^{5/2} \log C,$$

if n is taken sufficiently large. $\square$

Exercises

1. If $x_1, x_2, \ldots, x_n$ are linearly independent over $\mathbb{Q}$, show that the functions

$$e^{x_1 t}, e^{x_2 t}, \ldots, e^{x_n t}$$

are algebraically independent over the complex numbers. [Hint: use Exercise 1 of Chap. 4.]

2. Show that the functions t and e^t are algebraically independent over the field of complex numbers.

3. Show that at least one of $2^\pi, 3^\pi, 5^\pi$ is transcendental.

4. Let $\beta \in \mathbb{C}$ and suppose that there are three multiplicatively independent algebraic numbers $\alpha_1, \alpha_2, \alpha_3$ such that $\alpha_1^\beta, \alpha_2^\beta, \alpha_3^\beta$ are algebraic. Show that β is rational.

5. If p_1, p_2, p_3 are three distinct prime numbers such that p_1^x, p_2^x, p_3^x are integers, then show that x is a non-negative integer.

6. Imitate the above proof for the four exponential conjecture and find out where the proof breaks down.

7. Let $z \in \mathbb{C}$ with $|z| \in \mathbb{Q}$ and $e^{2\pi i z} \in \overline{\mathbb{Q}}$. Assuming the four exponential conjecture, deduce that $z \in \mathbb{Q}$.

Chapter 8

Estimates for Derivatives

In numerous transcendence proofs, it is convenient to estimate derivatives of polynomials evaluated at special points. To this end, we consider a more general setting.

We introduce some terminology. If P is a polynomial in several variables with algebraic coefficients, we will write size(P) for the maximum of the heights of its coefficients. Given two such polynomials, P and Q, with the latter having non-negative real coefficients, we will say that Q *dominates* P if the absolute value of the coefficient of each of the monomials in P is dominated by the corresponding coefficient of Q. We will write $P \prec Q$ if Q dominates P. It is easily verified that if $P_1 \prec Q_1$ and $P_2 \prec Q_2$, then $P_1 + P_2 \prec Q_1 + Q_2$ and $P_1 P_2 \prec Q_1 Q_2$. Moreover, if D_i is the derivative operator with respect to the i-th variable and $P \prec Q$, then $D_i P \prec D_i Q$. If the total degree of a polynomial P in n variables is r, then

$$P \prec \text{size}(P)(1 + x_1 + \cdots + x_n)^r.$$

We also need some facts about derivations. Recall that a *derivation* D of a ring R is a map $D : R \to R$ such that $D(x + y) = D(x) + D(y)$ and which satisfies $D(xy) = D(x)y + xD(y)$. Sometimes, we write Dx for $D(x)$ when the meaning is clear. For instance, if R is the polynomial ring $K[x_1, \ldots, x_n]$, then the partial derivative $\partial/\partial x_i$ is a derivation.

If R is an integral domain and K its quotient field, then a derivation D of R can be extended in the usual way by setting

$$D(u/v) = \frac{vD(u) - uD(v)}{v^2}.$$

If R is a ring with derivation D, then we can define a derivation on the polynomial ring $R[x_1, \ldots, x_n]$ by mapping the polynomial

$$f(x_1, \ldots, x_n) = \sum a_{i_1, \ldots, i_n} x_1^{i_1} \cdots x_n^{i_n}$$

M.R. Murty and P. Rath, *Transcendental Numbers*, DOI 10.1007/978-1-4939-0832-5_8, 31
© Springer Science+Business Media New York 2014

to

$$f^D := \sum D(a_{i_1,\dots,i_n}) x_1^{i_1} \cdots x_n^{i_n}.$$

It is easily verified that this is a derivation (see Exercise 2 below).

If $L = K[x_1, \dots, x_n]$, then the usual partial derivatives

$$D_i := \frac{\partial}{\partial x_i}$$

are derivations which are trivial on K. Conversely if D is a derivation of L which is trivial on K, then we can write it as a linear combination of the D_i's. Indeed, we have

$$D = \sum_i D(x_i) D_i$$

since any such derivation is determined by its values on the polynomials x_i. In fact, an easy induction shows that

$$D(x_i^m) = m x_i^{m-1} D(x_i)$$

and

$$D(x_1^{i_1} \cdots x_n^{i_n}) = \sum_{j=1}^{n} D_j(x_1^{i_1} \cdots x_n^{i_n}) D(x_j).$$

We apply these observations in the more familiar context of $\mathbb{C}(x_1, \dots, x_n)$. This shows that if $P_1, \dots, P_n$ are arbitrary polynomials, then there exists a unique derivation $\mathcal{D}^*$ such that $\mathcal{D}^*(x_i) = P_i$ which is trivial on $\mathbb{C}$.

Lemma 8.1 *Let K be an algebraic number field and $f_1, \dots, f_n$ be complex-valued functions. Let $w \in \mathbb{C}$ be such that the functions $f_1, \dots, f_n$ are holomorphic in a neighbourhood of w and that the derivative $D = d/dz$ maps the ring $K[f_1, \dots, f_n]$ into itself. Assume that $f_i(w) \in K$ for $1 \le i \le n$. Then there exists a number C_1 having the following property. Let $P(x_1, \dots, x_n)$ be a polynomial with coefficients in K and of degree $\deg(P) \le r$. If $f = P(f_1, \dots, f_n)$, then for all positive integers k,*

$$H(D^k f(w)) \le \operatorname{size}(P) k! C_1^{k+r}.$$

Moreover, the denominator of $D^k f(w)$ is bounded by $d(P) C_1^{k+r}$, where $d(P)$ is the denominator of the coefficients of P.

Proof. There exist polynomials $P_i(x_1, \dots, x_n)$ such that

$$D f_i = P_i(f_1, \dots, f_n).$$

Let δ be the maximum of their degrees. By our earlier remarks, there is a unique derivation $\mathcal{D}^*$ such that

$$\mathcal{D}^*(x_i) = P_i(x_1, \dots, x_n).$$

Then for any polynomial P, we have

$$\mathcal{D}^* P(x_1, \ldots, x_n) = \sum_{i=1}^{n} P_i(x_1, \ldots, x_n) \frac{\partial}{\partial x_i} P(x_1, \ldots, x_n).$$

The polynomial P is dominated by

$$\text{size}(P)(1 + x_1 + \cdots + x_n)^r$$

and so

$$\frac{\partial}{\partial x_i} P(x_1, \ldots, x_n)$$

is dominated by

$$\text{size}(P)r(1 + x_1 + \cdots + x_n)^r.$$

Now each P_i is dominated by

$$\text{size}(P_i)(1 + x_1 + \cdots + x_n)^\delta.$$

Thus, $\mathcal{D}^*(P)$ is dominated by

$$\text{size}(P)Cr(1 + x_1 + \cdots + x_n)^{r+\delta}$$

where

$$C = \sum_{i=1}^{n} \text{size}(P_i).$$

Now we argue similarly for $\mathcal{D}^{*2}(P)$. We have

$$\mathcal{D}^{*2} P(x_1, \ldots, x_n) = \sum_{i=1}^{n} P_i(x_1, \ldots, x_n) \frac{\partial}{\partial x_i} \mathcal{D}^* P(x_1, \ldots, x_n).$$

Since

$$\frac{\partial}{\partial x_i} \mathcal{D}^*(P) \prec \text{size}(P)r(r + \delta)C(1 + x_1 + \cdots + x_n)^{r+\delta},$$

we obtain

$$\mathcal{D}^{*2}(P) \prec \text{size}(P)r(r + \delta)C^2(1 + x_1 + \cdots + x_n)^{r+2\delta}.$$

Proceeding inductively, we see that

$$\mathcal{D}^{*k}(P) \prec \text{size}(P)C^k r(r + \delta) \cdots (r + (k-1)\delta)(1 + x_1 + \cdots x_n)^{r+k\delta}.$$

Observing that for $\delta > 0$ (for $\delta = 0$, the estimates are even easier),

$$r(r + \delta) \cdots (r + (k-1)\delta) \leq \delta^k(r + 1) \cdots (r + k) \leq \delta^k \frac{(r+k)!}{r!k!} k! \leq \delta^k 2^{r+k} k!,$$

we obtain an inequality of the form

$$\mathcal{D}^{*k}(P) \prec \text{size}(P) C_0^{r+k} k! (1 + x_1 + \cdots x_n)^{r+k\delta}.$$

If we plug in the values $f_i(w)$ for x_i in the above, we obtain a bound for $\mathcal{D}^{*k}(f(w))$ exactly of the form as required in our lemma. To prove the lemma, we observe that the map $x_i \mapsto f_i$ is a homomorphism from the ring $K[x_1, \ldots, x_n]$ to $K[f_1, \ldots, f_n]$ which takes the derivation D^* to D. Thus $D^{*k}(f(w)) = D^k f(w)$ and since all these numbers lie in the fixed number field K, the first assertion of the lemma follows. The second assertion about the denominator estimate follows in a similar inductive style. We leave this as an exercise to the reader. $\square$

Exercises

1. Let R be an integral domain and D a derivation of R. Show that the map D extends to the field of fractions of R by the definition:

$$D(x/y) = \frac{yD(x) - xD(y)}{y^2}.$$

2. Let D be a derivation on the ring R. Show that the map $f \mapsto f^D$ (defined in the beginning) is a derivation on the ring $R[x_1, \ldots, x_n]$.

3. If K is a field, show that the set of all derivations of K, denoted $\text{Der}(K)$, forms a vector space over K if we define

$$(D_1 + D_2)(x) := D_1(x) + D_2(x), \; (aD)(x) := aD(x),$$

for $D_1, D_2, D \in \text{Der}(K)$, and $a \in K$. Show further that $[D_1, D_2]$ $:= D_1 D_2 - D_2 D_1$ is again a derivation of K.

4. With notation as in the previous exercise, show that

$$[[D_1, D_2], D_3] + [[D_2, D_3], D_1] + [[D_3, D_1], D_2] = 0,$$

for any three derivations D_1, D_2, D_3 of K. (This is equivalent to saying that $\text{Der}(K)$ is a *Lie algebra*.)

5. Let D be a derivation on a field K and consider the function

$$L(x) = \frac{Dx}{x}, \qquad x \in K^{\times}.$$

Show that $L(xy) = L(x) + L(y)$. This map L is called the logarithmic derivative.

Chapter 9

The Schneider–Lang Theorem

In 1934, A.O. Gelfond and T. Schneider independently solved Hilbert's seventh problem. This problem predicted that if α and β are algebraic numbers with $\alpha \neq 0, 1$ and β irrational, then α^β is transcendental. In particular, the number $2^{\sqrt{2}}$ is transcendental as well as the number e^π, as is seen by taking $\beta = i$ and $\alpha = -1$. Another consequence of the theorem is the transcendence of numbers such as

$$\frac{\log \alpha}{\log \beta}$$

whenever $\log \alpha$ and $\log \beta$ are linearly independent over the rationals.

In 1962, Serge Lang derived a simple generalisation of the Schneider method and it is this result we will discuss here. In the subsequent chapters, we will derive further corollaries of the theorem.

Recall that an entire function f is said to be of *strict order* $\leq \rho$ if there is a positive constant C such that

$$|f(z)| \leq C^{R^\rho}$$

whenever $|z| \leq R$. A meromorphic function is said to be of *strict order* $\leq \rho$ if it is the quotient of two entire functions of strict order $\leq \rho$.

Theorem 9.1 (Schneider–Lang) *Let K be an algebraic number field. Let $f_1, \ldots, f_d$ be meromorphic functions of strict order $\leq \rho$ and assume that at least two of these functions are algebraically independent. Suppose further that the derivative $D = d/dz$ maps the ring $K[f_1, \ldots, f_d]$ into itself. If $w_1, \ldots, w_m$ are distinct complex numbers not among the poles of the f_i's such that $f_i(w_k) \in K$ for all $1 \leq i \leq d, 1 \leq k \leq m$, then $m \leq 4\rho[K : \mathbb{Q}]$.*

M.R. Murty and P. Rath, *Transcendental Numbers*, DOI 10.1007/978-1-4939-0832-5_9,   35
© Springer Science+Business Media New York 2014

Proof. Let f, g be two functions among $f_1, \ldots, f_d$ which are algebraically independent. Let

$$F = \sum_{i,j=1}^{r} a_{ij} f^i g^j.$$

We wish to select coefficients $a_{ij} \in \mathcal{O}_K$ such that

$$D^k F(w_\nu) = 0$$

for $1 \le \nu \le m$ and $0 \le k \le n-1$. This amounts to solving the linear system of mn equations in r^2 unknowns:

$$\sum_{i,j=1}^{r} a_{ij} D^k (f^i g^j)(w_\nu) = 0.$$

By hypothesis, the numbers

$$D^k(f^i g^j)(w_\nu) = \sum_{t=0}^{k} \binom{k}{t} D^t(f^i)(w_\nu) D^{k-t}(g^j)(w_\nu)$$

are algebraic and lie in K. By Lemma 8.1, we can estimate the size of our coefficients. Choosing $r^2 = 2mn$, Siegel's lemma assures that we can find the desired $a_{ij} \in \mathcal{O}_K$ with

$$H(a_{ij}) \le e^{n \log n + O(n+r)}.$$

Since f and g are algebraically independent over K, our function F is not identically zero. Let s be the smallest integer such that all the derivatives of F up to order $s-1$ vanish at the points $w_1, \ldots, w_m$ but such that $D^s F$ does not vanish at least at one of the w_ν, say w_1. Then $s \ge n$ and by Lemma 8.1, we have an estimate for

$$H(D^s F(w_1)) \le e^{s \log s + O(s)}.$$

We also know that it has denominator bounded by $e^{s \log s + O(s)}$. Since $D^s F(w_1) \ne 0$, from the height estimate we deduce that

$$|D^s(F(w_1))| \ge e^{-2[K:\mathbb{Q}]s \log s + s \log s + O(s)}.$$

On the other hand, we can deduce an upper bound for this quantity as follows. Let h be an entire function of order $\le \rho$ so that $h(w_1) \ne 0$ and both hf and hg are entire. Then

$$G(z) = \frac{h(z)^{2r} F(z)}{\prod_{\nu=1}^{m}(z-w_\nu)^s} \prod_{\nu=2}^{m}(w_1 - w_\nu)^s$$

is entire. Let us note that

$$\lim_{z \to w_1} \frac{G(z)}{h(z)^{2r}} = \frac{D^s(F(w_1))}{s!}.$$

because

$$F(z) = \frac{D^s(F(w_1))}{s!}(z - w_1)^s + \cdots$$

is its Laurent expansion about $z = w_1$. We note that to estimate $|D^s(F(w_1))|$, it suffices to estimate $G(z)$ on the circle $|z| = R$ which encloses w_1. For this, we apply the maximum modulus principle to G on the circle with radius $R = s^{1/2\rho}$ (suitably large) to obtain

$$\frac{|D^s(F(w_1))|}{s!} \ll C^{rR^\rho} R^{-ms}.$$

Since $R = s^{1/2\rho}$, we get an upper bound of

$$e^{-ms(\log s)/2\rho + s\log s + rs^{1/2}c_1}.$$

Recalling that $r = O(n^{1/2})$, we obtain a contradiction if $m > 4\rho[K : \mathbb{Q}]$. This completes the proof. $\square$

We now derive some important corollaries of this theorem.

Corollary 9.2 (Hermite–Lindemann Theorem) *Let α be a non-zero algebraic number. Then e^α is transcendental.*

Proof. Suppose not. Let K be the field generated by α and e^α over $\mathbb{Q}$. Let $f_1(z) = z$ and $f_2(z) = e^{\alpha z}$. Then the ring $K[f_1, f_2]$ is mapped into itself by the derivative map. Moreover, by Exercise 2 in Chap. 7, the two functions are algebraically independent. The theorem indicates that there are only finitely many complex numbers w such that $f_1(w), f_2(w) \in K$. But we may take the infinite set $w = 1, 2, 3, \ldots$ to derive a contradiction. $\square$

Corollary 9.3 (Gelfond–Schneider Theorem) *Let α, β be algebraic numbers with $\alpha \neq 0, 1$ and β irrational. Then α^β is transcendental.*

Proof. Suppose not. Let K be the field generated by $\alpha, \beta, \alpha^\beta$ over $\mathbb{Q}$. We apply the theorem to the two functions $f_1(z) = e^z$, $f_2(z) = e^{\beta z}$. Again, the derivative maps the ring $K[f_1, f_2]$ into itself. By Exercise 1 of Chap. 7, we conclude that f_1 and f_2 are algebraically independent. Thus f_1 and f_2 can take on values in K simultaneously at only a finite number of complex numbers. But this is a contradiction if we take $z = \log \alpha, 2\log \alpha, \ldots$. $\square$

In the subsequent chapters, we will discuss further applications of this important theorem to the theory of elliptic functions, abelian functions and modular functions.

Exercises

1. If P is a polynomial of degree d, determine the order of the function e^P.

2. Prove that if f is an entire function of order ρ, then its derivative f' also has order ρ.

3. Let α, β be algebraic numbers unequal to 0 or 1. Show that

$$\frac{\log \alpha}{\log \beta}$$

   is either rational or transcendental.

4. If α, β are non-zero algebraic numbers such that $\log \alpha$, $\log \beta$ are linearly independent over $\mathbb{Q}$, then show that they are linearly independent over $\overline{\mathbb{Q}}$.

Chapter 10

Elliptic Functions

Let ω_1, ω_2 be two complex numbers which are linearly independent over the reals. Let L be the lattice spanned by ω_1, ω_2. That is,

$$L = \{m\omega_1 + n\omega_2 : m, n \in \mathbb{Z}\}.$$

An *elliptic function* (relative to the lattice L) is a meromorphic function f on $\mathbb{C}$ (thus an analytic map $f : \mathbb{C} \to \mathbb{CP}_1$) which satisfies

$$f(z + \omega) = f(z)$$

for all $\omega \in L$ and $z \in \mathbb{C}$. The value of such a function can be determined by its value on the *fundamental parallelogram*:

$$D = \{s\omega_1 + t\omega_2 : 0 \leq s, t < 1\}.$$

Any translate of D is referred to as a *fundamental domain* for the elliptic functions relative to L. The set of all such elliptic functions (relative to L) forms a field and L is called the *period lattice* or the *lattice of periods*.

The *Weierstrass $\wp$-function* associated with L is defined by the series

$$\wp(z) = \frac{1}{z^2} + \sum_{\omega \in L'} \left\{ \frac{1}{(z - \omega)^2} - \frac{1}{\omega^2} \right\},$$

where L' denotes the set of non-zero periods. The associated *Eisenstein series of weight $2k$* is

$$G_{2k}(L) := \sum_{\omega \in L'} \omega^{-2k}.$$

Theorem 10.1 *Let L be a lattice in $\mathbb{C}$. The Eisenstein series G_{2k} is absolutely convergent for all $k > 1$. The Weierstrass $\wp$-function associated with L*

M.R. Murty and P. Rath, *Transcendental Numbers*, DOI 10.1007/978-1-4939-0832-5_10,   39
© Springer Science+Business Media New York 2014

converges absolutely and uniformly on every compact subset of $\mathbb{C}\backslash L$. It is a meromorphic elliptic function on $\mathbb{C}$ having a double pole at each point of L and no other poles.

Proof. It is easy to see that

$$\#\{\omega \in L : N \leq |\omega| < N + 1\} = O(N).$$

Hence

$$\sum_{\omega \in L: |\omega| \geq 1} \frac{1}{|\omega|^{2k}} \leq \sum_{N=1}^{\infty} \frac{1}{N^{2k}} \#\{\omega \in L: \quad N \leq |\omega| < N+1\} \ll \sum_{N=1}^{\infty} \frac{1}{N^{2k-1}},$$

from which the first assertion follows. To deal with the convergence of the second series, we split the series into two parts:

$$\sum_{0<|\omega|\leq 2|z|} + \sum_{|\omega|>2|z|}.$$

The first sum is a finite sum by our observation above. For the second sum, we note that

$$\left| \frac{1}{(z-\omega)^2} - \frac{1}{\omega^2} \right| = \left| \frac{z(2\omega-z)}{\omega^2(z-\omega)^2} \right| \leq \frac{10|z|}{|\omega|^3}$$

which converges by the first part of our theorem. Thus the defining series of the Weierstrass $\wp$-function converges absolutely and uniformly on every compact subset of $\mathbb{C}\backslash L$. This proves that $\wp$ is analytic in the region $\mathbb{C}\backslash L$. Further, we can compute its derivative and find that

$$\wp'(z) = -2 \sum_{\omega \in L} \frac{1}{(z-\omega)^3}$$

from which it is clear that $\wp'$ is an elliptic function. Thus for any fundamental period ω (i.e. there exists $\tau \in L$ such that ω and τ generate L),

$$\wp'(z+\omega) = \wp'(z)$$

and hence we obtain

$$\wp(z+\omega) = \wp(z) + c(\omega)$$

for some constant $c(\omega)$ which is independent of z. Putting $z = -\omega/2$ and noting that $\wp$ is an even function, we find that $c(\omega) = 0$. Finally the series representation clearly shows the location and multiplicities of the poles. This completes the proof. $\square$

The next theorem describes a fundamental algebraic relation between $\wp$ and $\wp'$.

Theorem 10.2 *The Laurent series for $\wp(z)$ about $z = 0$ is given by*

$$\wp(z) = \frac{1}{z^2} + \sum_{k=1}^{\infty} (2k + 1) G_{2k+2} z^{2k}.$$

Moreover, for all $z \in \mathbb{C}$, $z \notin L$ we have

$$\wp'(z)^2 = 4\wp(z)^3 - 60 G_4 \wp(z) - 140 G_6.$$

Proof. We begin by observing that

$$\sum_{n=0}^{\infty} z^n = \frac{1}{1-z}$$

for $|z| < 1$. Upon differentiating both sides, we find that

$$\sum_{n=0}^{\infty} n z^{n-1} = \frac{1}{(1-z)^2}.$$

Thus,

$$(1 - z)^{-2} - 1 = \sum_{n-1}^{\infty} (n+1) z^n,$$

a fact we will use below. Let $r = \min\{|w| : w \in L'\}$. Then, for $0 < |z| < r$, we can write

$$\frac{1}{(z-w)^2} - \frac{1}{w^2} = w^{-2}[(1 - z/w)^{-2} - 1] = \sum_{n=1}^{\infty} (n+1) z^n / w^{n+2}.$$

Summing both sides of this expression over $w \in L'$, we obtain

$$\wp(z) - \frac{1}{z^2} = \sum_{w \in L'} \sum_{n=1}^{\infty} (n+1) z^n / w^{n+2}.$$

Interchanging the summations on the right-hand side and noting that for odd $n \geq 1$, the sum

$$\sum_{w \in L'} w^{-n-2} = 0$$

(because both w and $-w$ are in L'), we obtain the first assertion of the theorem. To prove the second assertion, we differentiate the Laurent series to get

$$\wp'(z) = -2z^{-3} + 6G_4 z + 20 G_6 z^3 + \cdots.$$

Squaring this, we obtain

$$\wp'(z)^2 = 4z^{-6} - 24 G_4 z^{-2} - 80 G_6 + \cdots.$$

Cubing the $\wp$-function, we get

$$\begin{aligned}
\wp^3(z) &= \left(z^{-2} + 3G_4 z^2 + 5G_6 z^4 + \cdots\right)^3 \\
&= z^{-6} + 9G_4 z^{-2} + 15G_6 + \cdots.
\end{aligned}$$

Thus, the function

$$f(z) = \wp'(z)^2 - 4\wp(z)^3 + 60G_4\wp(z) + 140G_6$$

is holomorphic at $z = 0$ and vanishes there. Since $f(z + \omega) = f(z)$ for all $\omega \in L$, we have that f vanishes at all points of L. But it is also an elliptic function which is analytic outside of L. It follows that f is analytic on the fundamental parallelogram D and thus is an entire function. Since the closure of D is compact, f is a bounded entire function. By Liouville's theorem, f is constant. Since $f(0) = 0$, this constant must be zero. This completes the proof. $\square$

The preceding theorem shows that the points $(\wp(z), \wp'(z))$ for $z \in \mathbb{C}\backslash L$ lie on the curve defined by the equation

$$y^2 = 4x^3 - g_2 x - g_3,$$

where

$$g_2 = 60G_4, \quad g_3 = 140G_6.$$

The cubic polynomial on the right-hand side has a discriminant given by

$$\Delta = g_2^3 - 27g_3^2$$

which we shall show to be non-zero. Such curves are called *elliptic curves*.

We want to show that the converse is also true. Namely, given $(x, y) \in \mathbb{C}^2$ lying on the curve, we can find z such that $x = \wp(z)$ and $y = \wp'(z)$. Indeed, if the equation $\wp(z) - x = 0$ has no solution, then $1/(\wp - x)$ is an elliptic function which is holomorphic on L. By periodicity, we see that it is entire and bounded. By Liouville's theorem, it must be a constant, a contradiction since $\wp$ is not a constant function. Hence $y = \pm\wp'(z)$ and since $\wp'(z) = -\wp'(-z)$, we may adjust the sign of z appropriately so as to ensure that $(x, y) = (\wp(z), \wp'(z))$. This proves:

Theorem 10.3 *Let L be a lattice. Let g_2, g_3 be defined as above. Then all the complex solutions of the equation*

$$y^2 = 4x^3 - g_2 x - g_3$$

are given by $(\wp(z), \wp'(z))$ where $\wp$ is the Weierstrass $\wp$-function attached to L and z ranges over all the complex numbers in $\mathbb{C}\backslash L$.

Let us now prove the following elementary, but crucial lemma:

Lemma 10.4 *Let f be an elliptic function associated with the lattice L and let $\mathrm{res}_w f$ denote the residue of f at $z = w$. If D is a fundamental domain of f whose boundary ∂D does not contain any pole of f, then*

$$\sum_{w \in D} \mathrm{res}_w f = 0.$$

Further, if ord_w denotes the order of f at $z = w$ and ∂D does not contain a zero of f, then

$$\sum_{w \in D} \mathrm{ord}_w f = 0.$$

Proof. Let D be a fundamental domain whose boundary does not contain any pole of f. By Cauchy's theorem, we have

$$\int_{\partial D} f(z)dz = 2\pi i \sum_{w \in D} \mathrm{res}_w f.$$

The periodicity of f shows that the line integrals along the opposite sides of the parallelogram cancel. This proves the first assertion. The second one follows on applying the first assertion to the elliptic function $f'(z)/f(z)$. $\square$

There are two more related functions we will look at. The first is the *Weierstrass σ-function* attached to the lattice L and defined as

$$\sigma(z) := z \prod_{w \in L'} \left(1 - \frac{z}{\omega}\right) e^{z/\omega + z^2/2\omega^2}.$$

Since the series

$$\sum_{w \in L'} \omega^{-2-\epsilon}$$

is absolutely convergent for $\epsilon > 0$, *Weierstrass–Hadamard factorisation theory* for entire functions will immediately imply that σ is an entire function of order two. But without appealing to the general theory of entire functions, it is not difficult to carry out an explicit hands-on treatment of this function which we do.

If we formally take the logarithmic derivative of the σ function, we obtain the *Weierstrass ζ-function*:

$$\zeta(z) := \frac{\sigma'(z)}{\sigma(z)} = \frac{1}{z} + \sum_{w \in L'} \left[\frac{1}{z - \omega} + \frac{1}{\omega} + \frac{z}{\omega^2}\right].$$

The summand on the right can be written as

$$-\frac{1}{\omega(1 - z/\omega)} + \frac{1}{\omega} + \frac{z}{\omega^2} = -\sum_{k=2}^{\infty} \frac{z^k}{\omega^{k+1}}$$

for z in a suitable region. By our earlier remarks, we see that the series defining the function ζ converges absolutely and uniformly for any compact set in $\mathbb{C} \setminus L$. Thus ζ is an analytic function in $\mathbb{C} \setminus L$. Now exponentiating local primitives of this function, we see that σ is an entire function. If we differentiate ζ, we obtain

$$\zeta'(z) = -\frac{1}{z^2} - \sum_{w \in L'} \left[\frac{1}{(z-w)^2} - \frac{1}{w^2} \right] = -\wp(z).$$

It is easily seen that both σ and ζ are odd functions.

The Weierstrass σ-function has strict order ≤ 3. To see this, let z be a complex number of absolute value R. Then,

$$\log(1 - z/w) + z/w + z^2/2w^2 = -\sum_{k=3}^{\infty} \frac{z^k}{kw^k},$$

provided $|z| = R < |w|$. If $|w| > 2R$, then the sum converges absolutely since

$$\sum_{k=3}^{\infty} |z^k/w^k| \ll |z|^3/|w|^3.$$

Thus, the part of the product defining σ which is restricted to $|w| > 2R$ converges absolutely and its logarithm is $O(R^3)$. The part of the product over those w satisfying $|w| \leq 2R$ has $O(R^2)$ factors and each factor is $O(Re^{R^2})$ from which the assertion follows. From the product formula, we also see that σ is an entire function with simple zeros on L and at no other points.

Differentiating the function $\zeta(z+w) - \zeta(z)$, we get zero since the $\wp$-function is periodic. Thus there exists $\eta(w)$ so that

$$\zeta(z+w) = \zeta(z) + \eta(w).$$

It is clear that η is a $\mathbb{Z}$-linear function in w. Thus, $\eta(2w) = 2\eta(w)$. The notation $\eta_1 = \eta(w_1)$ and $\eta_2 = \eta(w_2)$ is standard and these are called *quasi-periods* of ζ. Thus ζ is not an elliptic function since it is not doubly periodic.

What about σ? From the preceding, we see that

$$\log \sigma(z+w) = \log \sigma(z) + \eta(w)z + c(w)$$

for some function c on the lattice. It is convenient to write this as

$$\frac{\sigma(z+w)}{\sigma(z)} = \psi(w)e^{\eta(w)(z+w/2)}$$

thereby defining $\psi(w)$. Suppose first that $w/2 \notin L$. Setting $z = -w/2$ above and using the fact that σ is odd, we see at once that $\psi(w) = -1$. On the other hand,

$$\frac{\sigma(z+2w)}{\sigma(z)} = \frac{\sigma(z+2w)}{\sigma(z+w)} \frac{\sigma(z+w)}{\sigma(z)}$$

and so by applying the functional equation twice and using the fact that $\eta(2\omega) = 2\eta(\omega)$, we get

$$\psi(2\omega) = \psi(\omega)^2.$$

In particular if $\omega/2 \in L$, we get

$$\psi(\omega) = \psi(\omega/2)^2.$$

Thus dividing by 2 until we get some element which is not twice a period, we get its value to be -1 which upon squaring becomes 1. This proves:

Theorem 10.5

$$\sigma(z + \omega) = \sigma(z)\psi(\omega)e^{\eta(\omega)(z+\omega/2)},$$

where $\psi(\omega) = 1$ if $\omega/2 \in L$ and -1 otherwise.

This theorem allows us to factor the $\wp$-function as a product of σ functions. Indeed, let us observe that for any $a \in \mathbb{C}$, we have

$$\frac{\sigma(z + a + \omega)}{\sigma(z + a)} = \psi(\omega)e^{\eta(\omega)(z+\omega/2)}e^{\eta(\omega)a}.$$

Noting that $\eta(\omega)a$ occurs linearly in the exponent, we see that if $a_1, \ldots, a_n$ and $b_1, \ldots, b_n$ are any two sets of complex numbers satisfying

$$\sum_{i=1}^{n} a_i = \sum_{i=1}^{n} b_i,$$

then the function

$$\prod_{i=1}^{n} \frac{\sigma(z - a_i)}{\sigma(z - b_i)}$$

is periodic with respect to the lattice L and hence an elliptic function. The converse is also true, namely that any elliptic function can be written as a product of above type. For example, we have for any $a \notin L$,

$$\wp(z) - \wp(a) = -\frac{\sigma(z + a)\sigma(z - a)}{\sigma^2(z)\sigma^2(a)}.$$

To see this, note that the left-hand side has zeros at $z = \pm a$ and all its translates by L and a double pole at $z = 0$. There are no other zeros or poles by Lemma 10.4. The right-hand side is an elliptic function by our earlier remarks with the zeros and poles of the same order and at the same places. Thus the quotient is entire and as its value is determined on the fundamental domain, it is bounded there. By Liouville's theorem, it is constant. Since $\sigma(z)/z$ tends to 1 as z tends to zero, we deduce that the constant must be 1 by multiplying both sides by z^2 and taking the limit as z tends to zero. This discussion along with Exercise 4 proves the following theorem.

Theorem 10.6 *Any elliptic function is expressible as a product of the form*

$$c \prod_{i=1}^{n} \frac{\sigma(z - a_i)}{\sigma(z - b_i)}$$

where c is a constant.

The subject of elliptic functions was developed in the nineteenth century by the works of Legendre, Gauss, Jacobi, Eisenstein, Kronecker and others. Today this field has grown naturally into the theory of modular forms, one of the most active branches of mathematics. We heartily recommend the delightful little book of Weil [131] which gives a wide-ranging historical perspective of this topic.

Exercises

1. Show that ζ and η are both odd functions.

2. Show that for any lattice L and $k \geq 2$,

$$G_{2k}(L) = P_k(G_4(L), G_6(L))$$

   where $P_k(x, y)$ is a polynomial with rational coefficients, independent of L.

3. Show that the Weierstrass $\wp$-function has strict order ≤ 3.

4. Prove that for any elliptic function f associated with a lattice L,

$$\sum_{w \in D} w \operatorname{ord}_w f \in L.$$

5. Fix a complex number c. Show that the equation $\wp(z) = c$ has exactly two solutions in the fundamental parallelogram. If u and v are these solutions, use the previous exercise to deduce that $u + v \in L$.

6. Prove that

$$\wp'(z) = -\frac{\sigma(2z)}{\sigma(z)^4}.$$

7. If f is an even elliptic function and u is a zero of order m, show that $-u$ is also a zero of order m. Prove the same assertion with "zero" replaced by "pole". Further, if $u = -u$ in $\mathbb{C}/L$, then m is even.

8. Prove that any even elliptic function f is a rational function in $\wp$. [Hint: By the previous exercise, pair up the zeros as $a_i, -a_i$ and the poles as

$b_i, -b_i$ in a fundamental domain, taking care when a pair is same mod L. Now consider the function

$$\frac{\prod_i(\wp(z) - \wp(a_i))}{\prod_i(\wp(z) - \wp(b_i))}$$

and show that it has the same zeros and poles as f.]

9. Conclude from the previous exercise that any elliptic function is a rational function in $\wp$ and $\wp'$.

Chapter 11

Transcendental Values of Elliptic Functions

The observation that points on a certain elliptic curve can be parametrised by the values of the $\wp$-function and its derivative allows us to deduce an important addition theorem for the $\wp$-function. Using Lemma 10.4, we will prove the following addition formula for the $\wp$-function.

Theorem 11.1 *For z_1, z_2 with z_1, z_2 and $z_1 \pm z_2 \notin L$, we have*

$$\wp(z_1 + z_2) = -\wp(z_1) - \wp(z_2) + \frac{1}{4}\left(\frac{\wp'(z_1) - \wp'(z_2)}{\wp(z_1) - \wp(z_2)}\right)^2.$$

When $z_1 = z_2 = z$, we have

$$\wp(2z) = -2\wp(z) + \frac{1}{4}\left(\frac{\wp''(z)}{\wp'(z)}\right)^2.$$

Proof. Let $(x_1, y_1) = (\wp(z_1), \wp'(z_1))$ and $(x_2, y_2) = (\wp(z_2), \wp'(z_2))$ be the corresponding points on the elliptic curve

$$y^2 = 4x^3 - g_2 x - g_3.$$

Let $y = ax + b$ be the line through these two points. Thus,

$$\wp'(z_1) = a\wp(z_1) + b, \quad \wp'(z_2) = a\wp(z_2) + b.$$

Now with a suitably chosen fundamental domain D (so that we can apply Lemma 10.4), we can ensure that the elliptic function

$$\wp'(u) - a\wp(u) - b$$

M.R. Murty and P. Rath, *Transcendental Numbers*, DOI 10.1007/978-1-4939-0832-5_11,   49
© Springer Science+Business Media New York 2014

has poles only at $z = 0$ in D and this pole has order 3. Thus by Lemma 10.4, it has three zeros in D (counting multiplicities). Working mod L, we already have two of these zeros, namely $u = z_1$ and $u = z_2$. Let $u = t$ be the third zero. Then

$$\wp'(t) = a\wp(t) + b$$

and by Exercise 4 in the previous chapter (which follows by integrating the function $z\frac{f'}{f}$), we have

$$z_1 + z_2 + t \in L.$$

In addition, we have

$$(\wp'(t))^2 = 4\wp(t)^3 - g_2\wp(t) - g_3.$$

We conclude that

$$(a\wp(t) + b)^2 = 4\wp(t)^3 - g_2\wp(t) - g_3.$$

By our analysis, the cubic equation

$$(ax + b)^2 = 4x^3 - g_2 x - g_3$$

has roots $x = \wp(z_1), \wp(z_2)$ and $x = \wp(t)$. Since the sum of the roots is $a^2/4$, we get

$$\wp(z_1) + \wp(z_2) + \wp(t) = a^2/4 = \frac{1}{4}\left(\frac{\wp'(z_1) - \wp'(z_2)}{\wp(z_1) - \wp(z_2)}\right)^2.$$

Since $-t = z_1 + z_2 \bmod(L)$ and $\wp(-t) = \wp(t)$, we deduce the assertion of the theorem. The second part is obtained by taking limits as z_1 tends to z_2. This completes the proof. $\square$

We are now ready to prove the following important application of the Schneider–Lang theorem.

Theorem 11.2 *Let L be a lattice and suppose that g_2, g_3 are algebraic. Then, for any algebraic $\alpha \notin L$, $\wp(\alpha)$ is transcendental.*

Proof. Suppose not. Then $\wp'(\alpha)$ is also algebraic since g_2, g_3 are algebraic. Since

$$\wp'(z)^2 = 4\wp(z)^3 - g_2\wp(z) - g_3,$$

we see upon differentiating the left-hand side and dividing by $\wp'(z)$ that

$$2\wp''(z) = 12\wp(z)^2 - g_2.$$

Let K be the algebraic number field generated by $g_2, g_3, \alpha, \wp(\alpha)$. We apply the Schneider–Lang theorem to the field K and the functions $z, \wp$ and $\wp'$. The derivative operator maps the polynomial ring generated by these functions into itself. These functions are of strict order ≤ 3 by the theory of the Weierstrass σ and ζ functions discussed in the previous chapter. Moreover the functions $f_1(z) = z$ and $f_2(z) = \wp(z)$ are algebraically independent. To see this, suppose they are algebraically dependent. Then there exists a polynomial $P \in \mathbb{C}[x, y]$ such that $P(z, \wp)$ is identically zero. If the x-degree of P is n and $n > 0$, we may write this relation as

$$z^n P_n(\wp(z)) + z^{n-1} P_{n-1}(\wp(z)) + \cdots + P_0(\wp(z)) = 0$$

for certain polynomials $P_0, \ldots, P_n$. If we take $z_0 \notin L$ with $P_n(\wp(z_0)) \neq 0$, we see from the periodicity of $\wp(z)$ that the polynomial

$$z^n P_n(\wp(z_0)) + z^{n-1} P_{n-1}(\wp(z_0)) \cdots + P_0(\wp(z_0))$$

has infinitely many zeros at $z_0 + \omega$ for $\omega \in L$. One can carry out a similar argument for the case when $n = 0$. Thus the functions f_1 and f_2 are algebraically independent. By the Schneider–Lang theorem, these functions and $\wp'$ take values in K simultaneously at only finitely many complex points. But this is not the case since by the addition formula for the $\wp$-function, these functions take values in K at all the points $n\alpha \notin L$ with $n = 1, 2, \ldots$. This contradicts the Schneider–Lang theorem. Thus $\wp(\alpha)$ must be transcendental. $\square$

Given a lattice L with invariants g_2 and g_3, let E be the elliptic curve $y^2 = 4x^3 - g_2 x - g_3$. As described in the previous chapter, the complex points of this curve are parametrized by the Weierstrass $\wp$-function of L.

Conversely, we shall see later that given an elliptic curve of the form $y^2 = 4x^3 - Ax - B$ (where $A^3 - 27B^2 \neq 0$ by definition), there exists a unique lattice L such that $g_2(L) = A$ and $g_3(L) = B$ and hence the complex points of this curve are parametrized by the Weierstrass $\wp$-function of L. The elements of L are referred to as *periods* of the given elliptic curve.

The previous theorem implies that any non-zero period of such an elliptic curve defined over $\overline{\mathbb{Q}}$ (i.e. g_2 and g_3 are algebraic) must be transcendental. To see this note that

$$\wp'(\omega_1/2) = -\wp'(-\omega_1/2)$$

since $\wp'$ is an odd function. But $\wp'$ is periodic with respect to L and so

$$\wp'(-\omega_1/2) = \wp'(-\omega_1/2 + \omega_1)$$

so that $\wp'(\omega_1/2) = 0$. The same reasoning shows that

$$\wp'(\omega_2/2) = \wp'((\omega_1 + \omega_2)/2) = 0.$$

From the fact that $(\wp(z), \wp'(z))$ are points on the elliptic curve

$$E: \quad y^2 = 4x^3 - g_2 x - g_3,$$

we immediately see that

$$\wp\left(\frac{\omega_1}{2}\right), \wp\left(\frac{\omega_2}{2}\right), \wp\left(\frac{\omega_1 + \omega_2}{2}\right)$$

are the zeros of the cubic equation

$$4x^3 - g_2 x - g_3 = 0.$$

These are called the *two-division points* of E. In particular it follows that for g_2, g_3 algebraic, all the two-division points are algebraic. Thus dividing any non-zero period by a suitable power of 2 and using the previous theorem, we immediately deduce the following fundamental result first proved by Schneider:

Theorem 11.3 *Any non-zero period of an elliptic curve*

$$y^2 = 4x^3 - g_2 x - g_3$$

with g_2 and g_3 algebraic is necessarily transcendental.

In other words, if L is the lattice with invariants $g_2(L) = g_2$ and $g_3(L) = g_3$, then $L \cap \overline{\mathbb{Q}} = 0$. This result could be viewed as the elliptic analogue of the transcendence of π since $2\pi i$ is a "period" of the exponential function e^z and $\wp$ is a higher dimensional generalisation of the exponential function in the sense that it is doubly periodic.

We end this chapter by noting the following result for future reference.

Proposition 11.4 *The numbers $\wp(\omega_1/2), \wp(\omega_2/2)$ and $\wp((\omega_1 + \omega_2)/2)$ are distinct.*

Proof. Suppose not. Let L be the lattice spanned by ω_1, ω_2 which are linearly independent over $\mathbb{R}$. Let us consider the function

$$f_1(z) = \wp(z) - \wp(\omega_1/2).$$

This has a double order zero at $z = \omega_1/2$ since $\wp'(\omega_1/2) = 0$. Since $\wp$ has a double order pole at $z = 0$ in a suitable translate of the fundamental parallelogram and no other poles, this accounts for all the zeros of f_1 by Lemma 10.4. It follows that any zero must be congruent to $\omega_1/2$ modulo L. If $\wp(\omega_2/2) = \wp(\omega_1/2)$, then we would have $\omega_1 \equiv \omega_2$ modulo L, contrary to their linear independence over $\mathbb{R}$. Thus $\wp(\omega_1/2)$ and $\wp(\omega_2/2)$ are distinct. A similar argument applies for the other two-division points. $\square$

We immediately deduce:

Proposition 11.5 *The discriminant*

$$g_2^3 - 27g_3^2 \neq 0.$$

Proof. It will be convenient to write

$$e_1 = \wp(\omega_1/2), \quad e_2 = \wp(\omega_2/2), \quad e_3 = \wp((\omega_1 + \omega_2)/2).$$

Then the discriminant of the cubic is

$$g_2^3 - 27g_3^2 = 16(e_1 - e_2)^2 (e_1 - e_3)^2 (e_2 - e_3)^2$$

and by the previous proposition, this is non-zero. $\square$

Exercises

1. If g_2, g_3 are algebraic, show that for any natural number n, the numbers $\wp(\omega_1/n)$ and $\wp(\omega_2/n)$ are algebraic numbers.

2. Let L be a lattice with g_2, g_3 algebraic. If $\wp(\alpha)$ is transcendental, show that $\wp(n\alpha)$ is transcendental for every natural number n with $n\alpha \notin L$.

3. With L as in the previous exercise, show that if $\wp(\alpha)$ is transcendental, then so is $\wp^{(n)}(\alpha)$ for every natural number n.

4. Show that the functions e^z and $\wp$ are algebraically independent.

5. Let g_2, g_3 be algebraic and w be any non-zero period of the elliptic curve $y^2 = 4x^3 - g_2 x - g_3$. Then show that w and π are linearly independent over $\overline{\mathbb{Q}}$.

Chapter 12

Periods and Quasiperiods

In the previous chapter, we proved that the fundamental periods ω_1, ω_2 of a Weierstrass $\wp$-function whose corresponding g_2, g_3 are algebraic are necessarily transcendental. A similar question arises for the nature of the associated quasi-periods η_1, η_2. We shall show that these are also transcendental whenever g_2 and g_3 are algebraic. To this end, we shall need the following lemmas. Let $\mathbb{H}$ denote the upper half-plane, i.e. the set of complex numbers z with $\Im(z) > 0$.

Lemma 12.1 (Legendre Relation) *If ω_1 and ω_2 are fundamental periods such that $\omega_1/\omega_2 \in \mathbb{H}$, then*

$$\omega_1\eta_2 - \omega_2\eta_1 = 2\pi i.$$

Proof. We integrate $\zeta(z)$ around a fundamental parallelogram D, shifted slightly so that the boundary does not contain a period. The only pole of ζ is at $z = 0$ with residue 1. Thus By Cauchy's theorem,

$$2\pi i = \int_{\partial D} \zeta(z)dz.$$

But $\zeta(z+\omega) = \zeta(z) + \eta(\omega)$ and hence the line integrals along the opposite sides of the parallelogram don't quite cancel, but give the required terms. $\square$

Lemma 12.2 *The functions $f_1(z) = \wp(z)$ and $f_3(z) = \alpha z + \beta\zeta(z)$, with α, β not both zero, are algebraically independent.*

Proof. We begin by observing the following facts (see exercises below):

$$\zeta(z_1 + z_2) = \zeta(z_1) + \zeta(z_2) + \frac{1}{2}\left(\frac{\wp'(z_1) - \wp'(z_2)}{\wp(z_1) - \wp(z_2)}\right)$$

M.R. Murty and P. Rath, *Transcendental Numbers*, DOI 10.1007/978-1-4939-0832-5_12, 55
© Springer Science+Business Media New York 2014

and

$$f_3(z_1 + z_2) = f_3(z_1) + f_3(z_2) + \frac{\beta}{2}\left(\frac{\wp'(z_1) - \wp'(z_2)}{\wp(z_1) - \wp(z_2)}\right).$$

Suppose now that f_1, f_3 are algebraically dependent and

$$f_3^n(z) + a_1(z)f_3(z)^{n-1} + \cdots + a_n(z) = 0 \qquad (12.1)$$

with $a_1, \ldots, a_n$ rational functions in $\wp(z)$. For rational integers c and d, we have

$$\begin{aligned} f_3(z + c\omega_1 + d\omega_2) &= \alpha(z + c\omega_1 + d\omega_2) + \beta\zeta(z + c\omega_1 + d\omega_2) \\ &= f_3(z) + \alpha(c\omega_1 + d\omega_2) + \beta(c\eta_1 + d\eta_2), \end{aligned}$$

by the quasi-periodicity of ζ. We claim that we can choose c, d such that

$$\theta := \alpha(c\omega_1 + d\omega_2) + \beta(c\eta_1 + d\eta_2) \neq 0.$$

Assume otherwise. Then choosing (c, d) equal to $(1, 0)$ and $(0, 1)$ respectively, we have

$$\alpha\omega_1 + \beta\eta_1 = 0 \quad \text{and} \quad \alpha\omega_2 + \beta\eta_2 = 0.$$

Multiplying the first equation by η_2 and the second one by η_1 and subtracting gives

$$\alpha(\omega_1\eta_2 - \omega_2\eta_1) = 0$$

which by Legendre's relation implies $\alpha = 0$. Similarly we deduce $\beta = 0$, contrary to hypothesis. Thus we can choose c, d such that $\theta \neq 0$. It follows by induction that

$$f_3(z + m(c\omega_1 + d\omega_2)) = f_3(z) + m\theta,$$

for every integer m. In (12.1), we replace z by $z + m(c\omega_1 + d\omega_2)$ to get

$$f_3^n(z + m(c\omega_1 + d\omega_2)) + a_{n-1}(z)f_3(z + m(c\omega_1 + d\omega_2))^{n-1} + \cdots + a_n(z) = 0,$$

since the a_i's are rational functions of $\wp$ which are periodic in ω_1 and ω_2. In the fundamental parallelogram, there are only finitely many values of z for which the functions $a_i(z)$ are not analytic. If we choose $z = z_0$ so that it is not one of these values, we obtain that the polynomial equation

$$x^n + a_{n-1}(z_0)x^{n-1} + \cdots + a_0(z_0) = 0$$

has infinitely many zeros:

$$f_3(z_0 + m(c\omega_1 + d\omega_2)) = f_3(z_0) + m\theta, \quad m = 1, 2, \ldots,$$

since $\theta \neq 0$. This is clearly a contradiction. $\square$

We note that the case $\beta = 0$ was established in an earlier chapter. Recall that ω is called a fundamental period of $\wp$ if there exists another period τ such that ω and τ generate the associated lattice L.

Theorem 12.3 *Let ω be a fundamental period of $\wp$ with g_2 and g_3 algebraic. Set $\eta_0 = 2\zeta(\omega/2)$. Then any linear combination of ω and η_0 with algebraic coefficients, not both coefficients zero, is transcendental.*

Proof. Suppose not. Suppose $\alpha\omega + \beta\eta_0$ is algebraic where α and β are algebraic numbers, not both zero. Consider the functions

$$f_1(z) = \wp(z), \ f_2(z) = \wp'(z), \ f_3(z) = \alpha z + \beta\zeta(z).$$

Let K be the algebraic number field generated by α, β and $\alpha\omega + \beta\eta_0$ together with the roots of the cubic equation

$$4x^3 - g_2 x - g_3 = 0.$$

Then the ring $K[f_1, f_2, f_3]$ is invariant under the differentiation map. We must check that at least two of the functions f_1, f_2, f_3 are algebraically independent. But this is clear from the previous lemma. We have already seen that f_1 and f_2 are quotients of entire functions of strict order ≤ 3 and the same is true for f_3. We will choose $z = (r + 1/2)\omega$ with r ranging over the integers. Since $\wp(\omega/2)$ is a root of the cubic equation

$$4x^3 - g_2 x - g_3 = 0,$$

we see that f_1 takes values in K at these points. We also see that f_2 vanishes at these points. Finally, since

$$\zeta(r\omega + \omega/2) = \zeta(\omega/2) + \eta(r\omega) = \eta_0/2 + r\eta(\omega),$$

$$f_3((r + 1/2)\omega) = \alpha\omega(r + 1/2) + \beta\eta_0/2 + \beta r\eta(\omega).$$

Since $\zeta(z + \omega) = \zeta(z) + \eta(\omega)$, putting $z = -\omega/2$ and using the fact that ζ is an odd function, we obtain

$$\zeta(\omega/2) = -\zeta(\omega/2) + \eta(\omega),$$

which proves that $\eta(\omega) = 2\zeta(\omega/2) = \eta_0$. So we obtain

$$f_3((r + 1/2)\omega) = (r + 1/2)(\alpha\omega + \beta\eta_0)$$

which lies in K by our assumption. We now have infinitely many complex numbers at which all these three functions simultaneously take values in K. This contradicts the Schneider–Lang theorem. $\square$

In particular, we have the following important result proved by Schneider;

Corollary 12.4 *If g_2, g_3 are algebraic, then any non-zero period or quasi-period is transcendental.*

Exercises

1. Prove that
$$\zeta(2z) = 2\zeta(z) + \frac{\wp''(z)}{2\wp'(z)}.$$

2. Show that if $\wp(\alpha)$ is algebraic, then $\zeta(2^n\alpha)$ is a polynomial in $\zeta(\alpha)$ with algebraic coefficients lying in a field of bounded degree over $\mathbb{Q}$.

3. Let L be a lattice with corresponding g_2, g_3 algebraic. If α is not a period, show that at least one of $\wp(\alpha), \zeta(\alpha)$ is transcendental.

4. Prove the addition formula for the ζ-function:
$$\zeta(z_1 + z_2) = \zeta(z_1) + \zeta(z_2) + \frac{1}{2}\left(\frac{\wp'(z_1) - \wp'(z_2)}{\wp(z_1) - \wp(z_2)}\right).$$

Chapter 13

Transcendental Values of Some Elliptic Integrals

In the case of trigonometric functions, we can rewrite the familiar identity

$$\sin^2 z + \cos^2 z = 1$$

as

$$y^2 + \left(\frac{dy}{dz}\right)^2 = 1$$

where $y(z) = \sin z$. We can retrieve the inverse function of sine by formally integrating

$$dz = \frac{dy}{\sqrt{1 - y^2}},$$

so that

$$\sin^{-1} z = \int_0^z \frac{dy}{\sqrt{1 - y^2}}.$$

The period of the sine function can also be retrieved from

$$2\pi = 4 \int_0^1 \frac{dy}{\sqrt{1 - y^2}}.$$

However, we should be cautious about this reasoning since $\sin^{-1} z$ is a multi-valued function and the integral may depend on the path taken from 0 to z. With this understanding, let us try to treat the inverse of the elliptic function $\wp(z)$ in a similar way. Indeed, we have

$$\frac{d\wp(z)}{dz} = \sqrt{4\wp(z)^3 - g_2\wp(z) - g_3}$$

M.R. Murty and P. Rath, *Transcendental Numbers*, DOI 10.1007/978-1-4939-0832-5_13,   59
© Springer Science+Business Media New York 2014

from which we intend to recover z as

$$z = \int \frac{dx}{\sqrt{4x^3 - g_2 x - g_3}}$$

upon setting $x = \wp(z)$.

As we mentioned before, it is an issue whether these integrals are well defined since they may depend on the path. To make these integrals well defined, we need to make branch cuts by appealing to the theory of Riemann surfaces. However, let us attempt to figure out the defect, namely the difference between integrals over two different paths. Let z_0 and z_1 be two fixed points and γ be a piecewise smooth path parametrized by $x = x(t)$, $0 \le t \le 1$, with $x(0) = z_0$ and $x(1) = z_1$. Suppose that the path does not pass through the any of the zeros of the polynomial $4x^3 - g_2 x - g_3$. Let $y = y(t)$ be a continuous path such that the points

$$(x(t), y(t)), \qquad 0 \le t \le 1$$

lie on the elliptic curve

$$E : y^2 = 4x^3 - g_2 x - g_3$$

associated with $\wp$. Let $y(0) = v_0$ and $y(1) = v_1$ be the end points of $y(t)$. Covering space theory for path lifting ensures that there exists a piecewise smooth path $u(t)$ such that

$$x(t) = \wp(u(t)), \qquad y(t) = \wp'(u(t)), \qquad 0 \le t \le 1.$$

Let w_0 and w_1 be the end points of $u(t)$, that is

$$\wp(w_0) = z_0, \ \wp'(w_0) = v_0 \quad \text{and} \quad \wp(w_1) = z_1, \ \wp'(w_1) = v_1.$$

Then we have,

$$I_\gamma = \int_\gamma \frac{dx}{y} = \int_0^1 \frac{x'(t)}{y(t)} dt = \int_0^1 \frac{\wp'(u(t))u'(t)}{\wp'(u(t))} dt = \int_0^1 u'(t) = w_1 - w_0.$$

Now let γ_1 be another path from z_0 to z_1 and suppose that the chosen y curve has the same beginning point v_0. Then it is clear that its terminal point is equal to $\pm v_1$. Let $I_{\gamma_1} = \int_{\gamma_1} \frac{dx}{y}$ be the integral with respect to this new path γ_1. Then it is not difficult to see that

$$I_{\gamma_1} = I_\gamma \ (\text{mod } L) \quad \text{or} \quad I_{\gamma_1} = -I_\gamma - 2w_0 \ (\text{mod } L),$$

according as the terminal point of the new y curve is v_1 or $-v_1$. The upshot of these discussions is that these integrals are to be interpreted up to the period lattice L.

These discussions also suggest a recipe to recover the periods of an elliptic curve $y^2 = f(x)$, namely by integrating dx/y along suitably chosen paths where $x = \wp(z)$.

For instance, if we integrate from $e_2 = \wp(\omega_2/2)$ to $e_3 = \wp((\omega_1 + \omega_2)/2)$, we get

$$\frac{\omega_1}{2} = \int_{e_2}^{e_3} \frac{dx}{\sqrt{4x^3 - g_2 x - g_3}}.$$

Similarly we have

$$\frac{\omega_2}{2} = \int_{e_1}^{e_3} \frac{dx}{\sqrt{4x^3 - g_2 x - g_3}}.$$

A similar comment can be made about quasi-periods. Indeed since

$$\zeta'(z) = -\wp(z),$$

we obtain

$$d\zeta = -\wp(z)dz = -\frac{\wp d\wp}{\sqrt{4\wp^3 - g_2\wp - g_3}}$$

which gives upon integration

$$-\eta_1/2 = -\zeta(\omega_1/2) = \int_{e_2}^{e_3} \frac{x\,dx}{\sqrt{4x^3 - g_2 x - g_3}}.$$

Many times it is more convenient to normalise the roots of $f(x)$ and reduce the curve to the form $E_\lambda : y^2 = x(x-1)(x-\lambda)$ with $\lambda \neq 0, 1$ satisfying $|\lambda| < 1$ and $|\lambda - 1| < 1$ (see [67], for instance). Then the following integrals

$$\int_{-\infty}^{0} \frac{dx}{\sqrt{x(x-1)(x-\lambda)}} \quad \text{and} \quad \int_{1}^{\infty} \frac{dx}{\sqrt{x(x-1)(x-\lambda)}}$$

determine a fundamental pair of periods for the curve E_λ. We shall come across such curves in a later chapter.

On the other hand, if the cubic polynomial $f(x) = 4x^3 - g_2 x - g_3$ in the Weierstrass form is defined over real numbers, then $f(x)$ becomes positive for x sufficiently large and for such x, the elliptic integral

$$\int_{x}^{\infty} \frac{dt}{\sqrt{f(t)}}$$

is easier to handle (see [46] for further properties of such integrals). In Chap. 16, we will explicitly evaluate some elliptic integrals of this type.

The subject of elliptic integrals constitutes an independent theme in mathematics. The reader may refer to the books [28, 46, 67, 114, 132] for more comprehensive treatment of the integrals considered in this chapter.

The elliptic integrals can be related to the problem of determining the circumference of an ellipse. To see this, let us consider the ellipse

$$\frac{x^2}{a^2} + \frac{y^2}{b^2} = 1,$$

with a, b real algebraic numbers and $0 < b < a$. We would like to calculate the perimeter of this ellipse. If we parametrize a curve in $\mathbb{R}^2$ by a map

$$t \mapsto (x(t), y(t)),$$

then as t goes from A to B, the length of the curve traversed is

$$\int_A^B \sqrt{x'(t)^2 + y'(t)^2} \, dt$$

which follows from elementary calculus. Since the ellipse can be parametrized by the map

$$t \mapsto (a \sin t, b \cos t)$$

for $0 \le t \le 2\pi$, we see that the perimeter of the ellipse is equal to

$$4 \int_0^{\pi/2} \sqrt{a^2 \cos^2 t + b^2 \sin^2 t} \, dt.$$

Putting $u = \sin t$, the integral becomes

$$\int_0^1 \sqrt{\frac{a^2 - (a^2 - b^2)u^2}{1 - u^2}} \, du.$$

In case $a = b$, this becomes $a\pi/2$. But when $a \neq b$, this is not an elementary function.

Let us set $k^2 = 1 - b^2/a^2$ so that the integral becomes

$$a \int_0^1 \sqrt{\frac{1 - k^2 u^2}{1 - u^2}} \, du.$$

If we put $t = 1 - k^2 u^2$, it is easy to see that the circumference is an algebraic multiple of

$$\int_{1-k^2}^1 \frac{t \, dt}{\sqrt{t(t-1)(t-(1-k^2))}}$$

which resembles a quasi-period. The curve

$$y^2 = t(t-1)(t-(1-k^2))$$

is not in the Weierstrass form, but can easily be put into that form by changing t to $t + (k^2 - 2)/3$. Making this change of variable shows that the circumference of an ellipse with algebraic major and minor axes is given by an algebraic linear combination of a period and a quasi-period of an elliptic curve defined over $\overline{\mathbb{Q}}$. Since it is non-zero, by the Schneider–Lang theorem, the circumference is transcendental. We shall see later that this circumference is related to some hypergeometric series.

In another set up, the regulator R_K of a number field K with positive unit rank measures the volume of the unit lattice of $\mathcal{O}_K$. But the transcendence of R_K is not known except for real quadratic fields. In a later chapter, we shall see that Schanuel's conjecture implies that R_K is transcendental.

Exercises

1. If $\alpha \neq 0$ is algebraic, show that $\tan \alpha$ is transcendental. Deduce that

$$\int_0^\alpha \frac{dx}{1+x^2}$$

   is transcendental for any non-zero algebraic α.

2. Show that if $0 < \alpha \leq 1$ and α is algebraic, then the integral

$$\int_0^\alpha \frac{dx}{\sqrt{1-x^2}}$$

   is transcendental.

3. Let α be algebraic and satisfy $0 \leq \alpha \leq 1$. Show that if $0 < k < 1$ and k is algebraic, then the integral

$$\int_0^\alpha \frac{x\,dx}{\sqrt{(1-x^2)(1-k^2x^2)}},$$

   is transcendental. What happens if $k = 1$? [Hint: put $x^2 = 1 - 1/t$.]

4. Prove that the integral

$$\int_1^\infty \frac{dx}{\sqrt{x^3 - 1}}$$

   is transcendental.

Chapter 14

The Modular Invariant

We begin with a discussion of an important result in complex analysis called the *uniformisation theorem*. We have shown how to associate a $\wp$-function with a given lattice L. Thus, $g_2 = g_2(L), g_3 = g_3(L)$ can be viewed as functions on the set of lattices. For a complex number z with imaginary part $\Im(z) > 0$, let L_z denote the lattice spanned by z and 1. We will denote the corresponding g_2, g_3 associated with L_z by $g_2(z)$ and $g_3(z)$. Thus,

$$g_2(z) = 60 \sum_{(m,n) \neq (0,0)} (mz + n)^{-4},$$

and

$$g_3(z) = 140 \sum_{(m,n) \neq (0,0)} (mz + n)^{-6}.$$

We set

$$\Delta(z) = g_2(z)^3 - 27 g_3(z)^2$$

which is the discriminant of the cubic defined by the corresponding Weierstrass equation. We first prove:

Lemma 14.1 $\Delta(z) \neq 0$.

Proof. This is equivalent to showing that the roots of the cubic equation

$$4x^3 - g_2(z)x - g_3(z) = 0$$

are distinct. But we have already seen this in Proposition 11.5 of Chap. 11. $\square$

M.R. Murty and P. Rath, *Transcendental Numbers*, DOI 10.1007/978-1-4939-0832-5_14, 65
© Springer Science+Business Media New York 2014

We now introduce the important j-function defined as

$$j(z) := 1728 \frac{g_2(z)^3}{g_2(z)^3 - 27g_3(z)^2}$$

which by the previous lemma is well defined for every z in the upper half-plane. We will use the modular invariant to address the following question: given two complex numbers A, B, with $A^3 - 27B^2 \neq 0$, does there exist a lattice L with $g_2(L) = A, g_3(L) = B$ so that its Weierstrass function $\wp$ satisfies the equation

$$\wp'(z)^2 = 4\wp(z)^3 - A\wp(z) - B?$$

As mentioned in the beginning, we can view g_2 and g_3 as functions on the set of lattices. Let L be a lattice spanned by two periods ω_1, ω_2. If we replace ω_1, ω_2 by $\lambda\omega_1, \lambda\omega_2$ with $\lambda \in \mathbb{C}^*$, we get another lattice denoted by λL. The g_2, g_3 of this new lattice λL get changed by a factor of λ^{-4} and λ^{-6} respectively and the corresponding elliptic curve is

$$y^2 = 4x^3 - \lambda^{-4}g_2 x - \lambda^{-6}g_3.$$

However if we change variables and replace x by $\lambda^{-2}x$ and y by $\lambda^{-3}y$, we find that we are reduced to the same Weierstrass equation as we started with.

Now suppose that L is a lattice generated by ω_1, ω_2 which are linearly independent over $\mathbb{R}$. Hence $\Im(\omega_1/\omega_2) \neq 0$ and by changing signs appropriately, we can arrange that this lies in the upper half-plane:

$$\mathbb{H} := \{z = x + iy : x, y \in \mathbb{R}, y > 0\}.$$

Let $SL_2(\mathbb{Z})$ be the group consisting of 2×2 matrices with integer entries and determinant 1, that is

$$SL_2(\mathbb{Z}) = \left\{ \sigma = \begin{pmatrix} a & b \\ c & d \end{pmatrix} \mid a, b, c, d \in \mathbb{Z}, ad - bc = 1 \right\}.$$

Then every such σ acts on a basis $[\omega_1, \omega_2]$ of L by sending it to

$$[a\omega_1 + b\omega_2, c\omega_1 + d\omega_2]$$

which generates the same lattice. Thus, the fundamental periods are not uniquely determined by the lattice. Conversely, two fundamental pairs $[\omega_1, \omega_2]$ and $[\omega_1', \omega_2']$ generate the same lattice only if they are congruent modulo the above action of $SL_2(\mathbb{Z})$.

The above action of $SL_2(\mathbb{Z})$ on the bases induces an action on the upper half-plane:

$$\begin{pmatrix} a & b \\ c & d \end{pmatrix} \cdot z := \frac{az + b}{cz + d}. \tag{14.1}$$

Recalling that for any $z \in \mathbb{H}$, $g_2(z)$ and $g_3(z)$ are precisely the g_2 and g_3 associated with the lattice L_z spanned by z and 1, we have

$$g_2\left(\frac{az + b}{cz + d}\right) = (cz + d)^4 g_2(z)$$

and

$$g_3 \left(\frac{az+b}{cz+d} \right) = (cz+d)^6 g_3(z).$$

Thus the modular function $j(z)$ satisfies

$$j \left(\frac{az+b}{cz+d} \right) = j(z)$$

and hence is invariant under the action of $SL_2(\mathbb{Z})$.

We now determine a fundamental domain for the action of $SL_2(\mathbb{Z})$ on $\mathbb{H}$. More precisely, we show that any z in the upper half-plane is equivalent to a point in the following region

$$\Im(z) > 0, \; -1/2 \leq \Re(z) \leq 1/2, \; |z| \geq 1.$$

This we show as follows. First note that for any $z \in \mathbb{H}$ and

$$\sigma = \begin{pmatrix} a & b \\ c & d \end{pmatrix} \in SL_2(\mathbb{Z}), \tag{14.2}$$

the imaginary part of $\sigma.z$ is given by

$$\Im(\sigma.z) = \frac{\Im(z)}{|cz+d|^2}. \tag{14.3}$$

Now let us isolate two distinguished elements T and S of $SL_2(\mathbb{Z})$ given by

$$T = \begin{pmatrix} 1 & 1 \\ 0 & 1 \end{pmatrix} \quad \text{and} \quad S = \begin{pmatrix} 0 & -1 \\ 1 & 0 \end{pmatrix}. \tag{14.4}$$

We see that $Tz = z+1$ and $Sz = -1/z$. Let $z \in \mathbb{H}$ be arbitrary. If $\Im(z) \geq 1$, repeated application of T ensures that z is equivalent to a point in the above-mentioned region. If $\Im(z) < 1$, we chose

$$\sigma = \begin{pmatrix} a & b \\ c & d \end{pmatrix} \in SL_2(\mathbb{Z}) \tag{14.5}$$

such that $|cz+d|$ is minimum and hence $\Im(\sigma \cdot z)$ is maximum (this is possible as $\mathbb{Z}$ is discrete). Let $w = \sigma \cdot z$. As before, applying T repeatedly to w ensures that w and hence z is equivalent to a point z_0 with $\Re(z_0) \in [-1/2, 1/2]$. Note that $\Im(w) = \Im(z_0)$. We claim that $|z_0| \geq 1$. For otherwise,

$$\Im(S \cdot z_0) = \frac{\Im(w)}{|z_0|^2} > \Im(w),$$

contradicting the maximality of $\Im(w)$. Hence any z in the upper half-plane is equivalent to a point in the region

$$\Im(z) > 0, \; -1/2 \leq \Re(z) \leq 1/2, \; |z| \geq 1.$$

One can show that if two points in this region are equivalent under the action of $SL_2(\mathbb{Z})$, then they lie on the boundary (see exercises below). We call this region the *standard fundamental domain* for the action of $SL_2(\mathbb{Z})$ on the upper half-plane. With a little more effort, one can deduce that $SL_2(\mathbb{Z})$ is generated by the matrices S and T.

Since the modular function $j(z)$ satisfies

$$j\left(\frac{az+b}{cz+d}\right) = j(z)$$

and hence is invariant under the action of $SL_2(\mathbb{Z})$, it defines a function on the quotient space $\mathbb{H}/SL_2(\mathbb{Z})$ to $\mathbb{C}$. We will prove later that the modular function takes every complex value (in fact, exactly once) on this quotient space. Note that the value zero implies the vanishing of $g_2(z)$. Assuming this fact, we can complete our proof of the uniformisation theorem as follows.

If we are given $(A, B) = (0, B)$ with B non-zero, we first choose z_0 so that $g_2(z_0) = 0$. Since $\Delta(z_0) \neq 0$, we have $g_3(z_0) \neq 0$. Now choosing λ such that

$$\lambda^{-6} g_3(z_0) = B,$$

the lattice $[\lambda z_0, \lambda]$ does the required job.

If $A \neq 0$, we proceed similarly. Let $a = B^2/A^3$. Observe that $a \neq 1/27$ since $A^3 - 27B^2 \neq 0$. Choose z_0 so that $j(z_0) = 1728/(1 - 27a)$. We can now multiply $g_2(z_0)$ and $g_3(z_0)$ appropriately to arrange $\lambda^{-6} g_3(z_0) = B$ and $\lambda^{-4} g_2(z_0) = A$. This completes the proof.

It remains to show that the j-function takes on every complex number precisely once. We begin by introducing the Bernoulli numbers. These are defined by the formal power series expansion:

$$\frac{x}{e^x - 1} = \sum_{k=0}^{\infty} B_k \frac{x^k}{k!}.$$

For example, $B_0 = 1, B_1 = -1/2, B_2 = 1/6, B_3 = 0$ and so on. One can show that $B_{2k+1} = 0$ for $k \geq 1$. Clearly these numbers are rational numbers. Our interest is to relate these to the values of the Riemann zeta function. We follow the exposition given in [109]. For $\Re(s) > 1$, the Riemann zeta function $\zeta(s)$ is defined as

$$\zeta(s) = \sum_{n=1}^{\infty} \frac{1}{n^s}.$$

This should *not* be confused with the Weierstrass ζ-function!

Following Euler, we begin by observing the product expansion for $\sin z$:

$$\sin z = z \prod_{n=1}^{\infty} \left(1 - \frac{z^2}{n^2 \pi^2}\right).$$

Taking logarithmic derivatives of both sides gives the following expansion for $z \notin \pi\mathbb{Z}$,

$$\cot z = \frac{1}{z} + \sum_{n=1}^{\infty} \frac{2z}{z^2 - n^2 \pi^2}. \tag{14.6}$$

Thus one has the following expansion around the origin,

$$z \cot z = 1 - 2 \sum_{n=1}^{\infty} \sum_{k=1}^{\infty} \frac{z^{2k}}{n^{2k} \pi^{2k}} = 1 - 2 \sum_{k=1}^{\infty} \zeta(2k) \frac{z^{2k}}{\pi^{2k}}. \qquad (14.7)$$

The left-hand side is

$$\frac{iz(e^{iz} + e^{-iz})}{e^{iz} - e^{-iz}}$$

which can be rewritten as

$$\frac{iz(e^{2iz} + 1)}{e^{2iz} - 1} = iz + \frac{2iz}{e^{2iz} - 1}.$$

This is easily seen to be

$$iz + \sum_{k=0}^{\infty} B_k \frac{(2iz)^k}{k!}.$$

We immediately deduce:

Theorem 14.2 *If $\zeta(s)$ is the Riemann zeta function, then for $k \geq 1$,*

$$\zeta(2k) = -B_{2k} \frac{(2\pi i)^{2k}}{2(2k)!}.$$

In particular, each of these values is a transcendental number.

It is interesting to note that this derivation also shows directly that $B_{2k+1} = 0$ for $k \geq 1$ and that $(-1)^{k+1} B_{2k} > 0$. In particular, we deduce from Theorem 14.2 that

$$\zeta(2) = \frac{\pi^2}{6}, \quad \zeta(4) = \frac{\pi^4}{90} \quad \text{and} \quad \zeta(6) = \frac{\pi^6}{945}.$$

We would like to relate these observations to the Eisenstein series G_4, G_6 introduced earlier. From Eq. (14.6), we see that

$$\pi \cot \pi z = \frac{1}{z} + \sum_{m=1}^{\infty} \left(\frac{1}{z+m} + \frac{1}{z-m} \right).$$

On the other hand, writing $q = e^{2\pi i z}$, we have

$$\pi \cot \pi z = \pi \frac{\cos \pi z}{\sin \pi z} = i\pi \frac{q+1}{q-1} = i\pi - \frac{2\pi i}{1-q} = i\pi - 2\pi i \sum_{n=0}^{\infty} q^n.$$

Comparing this with (14.6), we obtain

$$\frac{1}{z} + \sum_{m=1}^{\infty} \left(\frac{1}{z+m} + \frac{1}{z-m} \right) = i\pi - 2\pi i \sum_{n=0}^{\infty} q^n.$$

By successive differentiations of the above, we get the following formula (valid for $k \geq 2$):

Theorem 14.3

$$\sum_{m=-\infty}^{\infty} \frac{1}{(m+z)^k} = \frac{1}{(k-1)!}(-2\pi i)^k \sum_{n=1}^{\infty} n^{k-1} q^n.$$

Using this result, we will obtain the following expansion of the Eisenstein series:

$$G_{2k}(z) := \sum_{(m,n) \neq (0,0)} (mz+n)^{-2k}.$$

Indeed, separating out $m=0$ from $m \neq 0$, we get

$$G_{2k}(z) = 2\zeta(2k) + 2 \sum_{m=1}^{\infty} \sum_{n \in \mathbb{Z}} (mz+n)^{-2k},$$

and using the previous theorem with z replaced by mz, and k replaced by $2k$,

$$G_{2k}(z) = 2\zeta(2k) + \frac{2(-2\pi i)^{2k}}{(2k-1)!} \sum_{d=1}^{\infty} \sum_{a=1}^{\infty} d^{2k-1} q^{ad}.$$

If we define the function

$$\sigma_k(n) := \sum_{d|n} d^k,$$

we may write this expansion as follows.

Theorem 14.4

$$G_{2k}(z) = 2\zeta(2k) + \frac{2(2\pi i)^{2k}}{(2k-1)!} \sum_{n=1}^{\infty} \sigma_{2k-1}(n) q^n, \quad q = e^{2\pi i z}.$$

This is the Taylor expansion of G_{2k} at $i\infty$, once the one-point compactification of $\mathbb{H}$ is endowed with a suitable Riemann surface structure.

We would like to relate this to g_2 and g_3 defined earlier. Indeed, an easy calculation shows that

$$\Delta(z) = (2\pi)^{12}(q - 24q^2 + 252q^3 - 1472q^4 + \cdots).$$

The coefficients of the power series in the brackets define the *Ramanujan τ-function*. Furthermore, it can be shown that

$$\sum_{n=1}^{\infty} \tau(n) q^n = (q - 24q^2 + 252q^3 - 1472q^4 + \cdots) = q \prod_{n=1}^{\infty} (1 - q^n)^{24}.$$

The Ramanujan τ-function has played a central role in the development of modern theory of automorphic forms. It is an important conjecture, due to Lehmer, that $\tau(n)$ is never equal to zero.

These expansions for G_{2k} and Δ are enough to prove that the modular function j takes every complex value precisely once. For this, we view j as a meromorphic function on the compactified Riemann surface $\widehat{\mathbb{H}/\Gamma}$ where $\Gamma = SL_2(\mathbb{Z})$. We refer to [112] for a detailed topological as well as analytic description of this space. The map

$$z \to q = e^{2\pi i z}$$

gives the local parameter at $i\infty$. In other words, for any Γ-invariant analytic function f on $\mathbb{H}$, we first express f as a function of q by composing with the local inverse of the map $z \to q$. This defines an analytic function on the punctured unit disc $0 < |q| < 1$. The behaviour of this function at the origin determines the behaviour of f at $i\infty$. Recalling the definition of j and using the q expansions for Δ and G_4, we have the following q expansion for the j function:

$$j(z) = \frac{1}{q} + 744 + 196884q + \cdots .$$

Thus the j function has a simple pole at $i\infty$. Since a meromorphic function on a compact Riemann surface has an equal number of zeros as poles, we see that the equation $j(z) = c$ has exactly one solution since j has only a simple pole at $i\infty$. In other words, the j function defines an analytic isomorphism between the compact Riemann surface $\widehat{\mathbb{H}/\Gamma}$ and the Riemann sphere $\mathbb{CP}_1$.

Finally, let L and L' be two lattices with the same invariants, i.e. $g_2(L) = g_2(L')$ and $g_3(L) = g_3(L')$. Then their respective Weierstrass $\wp$-functions have the same Laurent expansion at the origin (see Exercise 2 of Chap. 10) and therefore must agree everywhere. Thus, they must have the same set of poles and hence $L = L'$.

This completes the proof of the uniformisation theorem and we record this as:

Theorem 14.5 *Let A, B be two complex numbers such that $A^3 - 27B^2 \neq 0$. There exists a unique lattice L with $g_2(L) = A, g_3(L) = B$ and an associated $\wp$-function that satisfies*

$$\wp'(z)^2 = 4\wp(z)^3 - A\wp(z) - B.$$

As indicated before, we shall call L to be the *period lattice* of the elliptic curve $y^2 = 4x^3 - Ax - B$.

We now define the j-invariant associated with the elliptic curve

$$E: \quad y^2 = 4x^3 - Ax - B$$

as $j(E) = 1728A^3/(A^3 - 27B^2)$.

Suppose that we are given two period lattices L and L^* with corresponding Weierstrass functions $\wp$ and $\wp^*$, as well as corresponding g_2, g_3 and g_2^*, g_3^*. We would like to determine when the corresponding elliptic curves are isomorphic. That is, when is there an analytic isomorphism

$$\phi : \mathbb{C}/L \xrightarrow{\sim} \mathbb{C}/M$$

where ϕ is also a group homomorphism?

Let us first try to characterise analytic maps between such tori. Since the natural maps from $\mathbb{C}$ to the quotients $\mathbb{C}/L$ and $\mathbb{C}/M$ are universal covering maps, any such analytic map

$$\phi : \mathbb{C}/L \to \mathbb{C}/M$$

lifts to an analytic function $\tilde{\phi} : \mathbb{C} \to \mathbb{C}$. Now for any $\omega \in L$, consider the function

$$f_\omega(z) = \tilde{\phi}(z + \omega) - \tilde{\phi}(z).$$

This analytic function is mapped into M and hence is constant. Differentiating, we see that $\tilde{\phi}'$ is an analytic elliptic function with respect to the lattice L and therefore is also constant. This implies that $\tilde{\phi}(z)$ is of the form $az + b$ for some $a, b \in \mathbb{C}$. Since $\tilde{\phi}$ is the lift of the map $\phi : \mathbb{C}/L \to \mathbb{C}/M$, we see that $aL \subset M$. In other words, every analytic map

$$\phi : \mathbb{C}/L \to \mathbb{C}/M$$

is necessarily of the form

$$\phi(z + L) = az + b + M$$

where $aL \subset M$. Clearly, ϕ is invertible if and only if $aL = M$. Finally, if we require ϕ to be a group homomorphism, then $\phi(0) = 0$ and hence

$$\phi(z + L) = az + M.$$

We record these observation in the following theorem.

Theorem 14.6 *If $\phi : \mathbb{C}/L \to \mathbb{C}/M$ is an analytic homomorphism, then $\phi(z + L) = \alpha z + M$ for some complex number α and $\alpha L \subseteq M$. In particular, two lattices L and M give rise to isomorphic elliptic curves if and only if there is a complex number α such that $\alpha L = M$.*

Any non-zero analytic homomorphism between elliptic curves is called an *isogeny*. Further, we say two lattices L, M are *homothetic* if $\alpha L = M$ for some complex number α. Clearly this is an equivalence relation. The above theorem says that there is a one-to-one correspondence between isomorphism classes of elliptic curves over $\mathbb{C}$ and homothety classes of lattices of rank 2 over $\mathbb{R}$.

From this theorem, we will deduce that two elliptic curves are isomorphic if and only if their j-invariants are equal. One way is obvious, namely if E_1 and E_2 are isomorphic, then their corresponding lattices are homothetic and hence $j(E_1) = j(E_2)$.

To establish the converse, recall that for a given lattice L, there exists a basis $[\omega_1, \omega_2]$ with $\tau = \omega_1/\omega_2$ in the upper half-plane $\mathbb{H}$. Thus any lattice is homothetic to a lattice of the form $L_\tau = \mathbb{Z}\tau + \mathbb{Z}$ where $\tau \in \mathbb{H}$. Further, for any two points τ and τ' in $\mathbb{H}$, L_τ and $L_{\tau'}$ are homothetic if and only if there exists a $\sigma \in SL_2(\mathbb{Z})$ such that $\sigma.\tau = \tau'$. This means that elements in the quotient $\mathbb{H}/SL_2(\mathbb{Z})$ can be identified with the set of lattices up to homothety.

Now let E_1, E_2 be two elliptic curves with $j(E_1) = j(E_2)$. Let their corresponding lattices be L and M respectively. Suppose that they have the same j invariant. By the above theorem, E_1, E_2 are isomorphic if L and M are homothetic. Recall that for the lattice L, there is a unique point τ in $\mathbb{H}/SL_2(\mathbb{Z})$ such that L is homothetic to $L_\tau = \mathbb{Z}\tau + \mathbb{Z}$. Let τ' be such point in $\mathbb{H}/SL_2(\mathbb{Z})$ such that M is homothetic to $L_{\tau'}$. But since $j(E_1) = j(E_2)$, we have $j(\tau) = j(\tau')$. By the injectivity of j on $\mathbb{H}/SL_2(\mathbb{Z})$, we deduce that $\tau = \tau'$. Thus L and M are homothetic and hence E_1 and E_2 are isomorphic. This proves:

Theorem 14.7 *Two elliptic curves E_1 and E_2 are isomorphic over $\mathbb{C}$ if and only if $j(E_1) = j(E_2)$.*

Theorem 14.6 allows us to study the endomorphism rings of elliptic curves. Let E be an elliptic curve with period lattice given by $L = [\omega_1, \omega_2]$ with $\tau = \omega_1/\omega_2 \in \mathbb{H}$. Then as observed before, all analytic homomorphisms

$$\phi : \mathbb{C}/L \to \mathbb{C}/L$$

are of the form $\phi(z+L) = \alpha z + L$ for some α satisfying $\alpha L \subseteq L$. In other words, each endomorphism corresponds to a complex number α satisfying

$$\alpha\omega_1 = a\omega_1 + b\omega_2, \quad \alpha\omega_2 = c\omega_1 + d\omega_2$$

for integers a, b, c, d. In particular, α is an eigenvalue of a matrix with integer entries. Thus it is an algebraic integer of degree at most two over $\mathbb{Q}$. Clearly $\text{End}(E)$ contains an isomorphic copy of $\mathbb{Z}$ since the maps $z \mapsto nz$ have the property that $nL \subseteq L$.

If $\text{End}(E)$ is larger than $\mathbb{Z}$, then let $\alpha \in \text{End}(E)$ be such that $\alpha \notin \mathbb{Z}$. Working with the homothetic lattice $L_\tau = [\tau, 1]$, the above equations read

$$\alpha\tau = a\tau + b, \quad \alpha = c\tau + d.$$

This implies

$$\tau(c\tau + d) = (a\tau + b).$$

Since $\alpha \notin \mathbb{Z}$, we have $c \neq 0$. This means that τ is an algebraic number of degree 2 and $\mathbb{Q}(\alpha) = \mathbb{Q}(\tau)$. Note that τ lies in the upper half-plane and therefore generates an imaginary quadratic field. Hence the ring of endomorphisms of E can be identified with a subring of the ring of integers of an imaginary quadratic field.

Thus we may partition elliptic curves into two groups, those whose endomorphism ring is isomorphic to $\mathbb{Z}$ and those for which it is larger. In the second case, as we observed above, the endomorphism ring is a subring of an imaginary quadratic field $\mathbb{Q}(\tau)$. Furthermore, such a subring is an order in the imaginary quadratic field $k = \mathbb{Q}(\tau)$. We recall that an *order* $\mathcal{O}$ in a number field K is a subring (with unity) of the ring of integers $\mathcal{O}_K$ which also contains a $\mathbb{Q}$-basis of K. In such case, we say E is a CM curve (CM standing for complex multiplication) and in the former case, we say the curve is non-CM. Points τ in the standard fundamental domain D for which $\mathbb{Q}(\tau)$ is an imaginary quadratic field are sometimes called *CM points* for obvious reasons.   One can be a bit more precise. If $\mathcal{O}$ is such an order in k, then $\mathcal{O} = \mathbb{Z} + f\mathcal{O}_k$ for some positive integer f. This integer f is called the *conductor* of $\mathcal{O}$.

Now for any order $\mathcal{O}$ in a number field K, the group of invertible fractional ideals of $\mathcal{O}$ modulo the subgroup of principal ideals forms a finite abelian group. This is called the *Picard group* of $\mathcal{O}$. For instance, if $\mathcal{O}$ is equal to the ring of integers $\mathcal{O}_K$, then its Picard group is the usual ideal class group of K.

Let $\mathcal{O}$ be an order in an imaginary quadratic field k. Then it is known that the set of isomorphism classes of elliptic curves E over $\mathbb{C}$ whose endomorphism ring $\text{End}(E)$ is equal to $\mathcal{O}$ is in bijection with the Picard group of $\mathcal{O}$ (see [36], for instance).

Exercises

1. Show that $SL_2(\mathbb{Z})$ is generated by the matrices
$$\begin{pmatrix} 1 & 1 \\ 0 & 1 \end{pmatrix} \quad \text{and} \quad \begin{pmatrix} 0 & -1 \\ 1 & 0 \end{pmatrix}.$$

2. Prove that any two interior points of the region
$$\mathcal{D} = \{\Im(z) > 0, \quad -1/2 \le \Re(z) \le 1/2, \quad |z| \ge 1\},$$
are inequivalent under the action of $SL_2(\mathbb{Z})$.

3. Justify the interchange of summations in formula (14.7).

4. Let $\mathcal{D}^*$ be the compactified upper half-plane modulo $SL_2(\mathbb{Z})$. Show that any meromorphic function on $\mathcal{D}^*$ has only a finite number of zeros and poles.

5. Prove that any meromorphic function f on the upper half-plane satisfying
$$f\left(\frac{az + b}{cz + d}\right) = f(z)$$
for all
$$\begin{pmatrix} a & b \\ c & d \end{pmatrix} \in SL_2(\mathbb{Z})$$
is a rational function in j.

6. If $\mathcal{O}$ is an order in an imaginary quadratic field k, then show that $\mathcal{O} = \mathbb{Z} + f\mathcal{O}_k$ for some positive integer f.

Chapter 15

Transcendental Values of the j-Function

Let L and M be two lattices with corresponding Weierstrass functions $\wp$ and $\wp^*$. We begin by showing that if $\wp$ and $\wp^*$ are algebraically dependent, then there is a natural number m such $mM \subseteq L$. Indeed suppose that $\wp$ and $\wp^*$ are as above and there is a polynomial $P(x, y) \in \mathbb{C}[x, y]$ such that $P(\wp, \wp^*) = 0$. Then for some rational functions $a_i(x)$ and some natural number n, we have

$$\wp(z)^n + a_{n-1}(\wp^*(z))\wp(z)^{n-1} + \cdots + a_0(\wp^*(z)) = 0.$$

Choose $z_0 \in \mathbb{C}$ so that $\wp^*(z_0)$ is not a pole of the $a_i(z)$ for $0 \le i \le n-1$. This can be done since the $a_i(z)$ are rational functions and so there are only finitely many values to avoid in a fundamental domain. Then

$$\wp(z_0)^n + a_{n-1}(\wp^*(z_0))\wp(z_0)^{n-1} + \cdots + a_0(\wp^*(z_0)) = 0.$$

If $\omega^* \in M$, then we get

$$\wp(z_0 + \omega^*)^n + a_{n-1}(\wp^*(z_0))\wp(z_0 + \omega^*)^{n-1} + \cdots + a_0(\wp^*(z_0)) = 0.$$

Thus $\wp(z_0 + \omega^*)$, as ω^* ranges over elements of M, are also zeros of the polynomial

$$z^n + a_{n-1}(\wp^*(z_0))z^{n-1} + \cdots + a_0(\wp^*(z_0)) = 0.$$

In particular, this is true of multiples of ω_1^* and ω_2^*. We therefore get infinitely many roots of the above polynomial equation unless $mM \subseteq L$ for some positive natural number m. We record these observations in the following.

M.R. Murty and P. Rath, *Transcendental Numbers*, DOI 10.1007/978-1-4939-0832-5_15, 75
© Springer Science+Business Media New York 2014

Theorem 15.1 *Let L and M be two lattices with corresponding Weierstrass functions $\wp$ and $\wp^*$. Then $\wp$ and $\wp^*$ are algebraically dependent if and only if there is some natural number $m > 0$ such that $mM \subseteq L$.*

Proof. We have already established the "only if" part of this assertion. For the converse, suppose that $mM \subseteq L$. Then $\wp(mz)$ is periodic with respect to M. Thus it is an even elliptic function with respect to M. As noted in an earlier chapter, this means that it is a rational function in $\wp^*$. On the other hand, $\wp(mz)$ is also a rational function in $\wp(z)$. Thus, $\wp$ and $\wp^*$ are algebraically dependent. This completes the proof. $\square$

We are now ready to prove the following theorems due to Schneider.

Theorem 15.2 *Suppose that $\wp$ and $\wp^*$ have corresponding g_2, g_3 and g_2^*, g_3^* algebraic and assume that they are algebraically independent. Then*

$$\wp(z), \wp^*(z)$$

cannot take algebraic values simultaneously.

Proof. Let us suppose that z_0 is such that both $\wp(z)$ and $\wp^*(z)$ are algebraic. Let K be the field generated by

$$g_2, g_3, g_2^*, g_3^*, \wp(z_0), \wp'(z_0), \wp^*(z_0), \wp^{*'}(z_0).$$

We apply the Schneider–Lang theorem with the functions

$$\wp, \wp', \wp^*, \wp^{*'}.$$

By hypothesis, $\wp$ and $\wp^*$ are algebraically independent. The Schneider–Lang theorem says that there are only finitely many complex numbers for which all these functions take algebraic values in K. But this is a contradiction since they take algebraic values in K for the points nz_0 as n runs over an infinite family of integers. This completes the proof. $\square$

Theorem 15.3 *If α is an algebraic number in the upper half-plane which is not a quadratic irrational, then $j(\alpha)$ is transcendental.*

Proof. Let ω_1, ω_2 be such that $\omega_1/\omega_2 = \alpha$. Suppose that $j(\alpha)$ is algebraic. Replacing ω_1, ω_2 by $\lambda\omega_1, \lambda\omega_2$ we can arrange $g_2 = 1$ if $j(\alpha) \neq 0$ and $g_3 = 1$ if $j(\alpha) = 0$. In this way, we can arrange g_2, g_3 algebraic. Thus without loss of generality, we may work with a lattice $L = [\omega_1, \omega_2]$ with algebraic invariants such that $\omega_1/\omega_2 = \alpha$. Now let $\omega_1^* = \alpha\omega_1, \omega_2^* = \alpha\omega_2$ and denote by M the lattice spanned by ω_1^*, ω_2^*. Thus, $g_2^* = \alpha^{-4}g_2$, and $g_3^* = \alpha^{-6}g_3$. Also,

$$\wp^*(\alpha z) = \alpha^{-2}\wp(z).$$

In particular, setting $z = \omega_2/2$ gives

$$\wp^*(\omega_1/2) = \alpha^{-2}\wp(\omega_2/2)$$

so that both $\wp(\omega_1/2)$ and $\wp^*(\omega_1/2)$ are algebraic. This contradicts the previous theorem unless $\wp$ and $\wp^*$ are algebraically dependent. By Theorem 15.1, this means that there is a natural number m such that $mM \subseteq L$. In particular,

$$\omega_1^* = a\omega_1 + b\omega_2, \quad \omega_2^* = c\omega_1 + d\omega_2$$

for some rational numbers a, b, c, d such that $ad - bc \neq 0$. [This is because, ω_1, ω_2 are linearly independent over $\mathbb{R}$, as well as ω_1^*, ω_2^*.] Thus

$$\alpha \begin{pmatrix} \omega_1 \\ \omega_2 \end{pmatrix} = \begin{pmatrix} a & b \\ c & d \end{pmatrix} \begin{pmatrix} \omega_1 \\ \omega_2 \end{pmatrix}$$

which means that α is an eigenvalue of the matrix

$$\begin{pmatrix} a & b \\ c & d \end{pmatrix}.$$

Since a, b, c, d are rational numbers, this means that α is a quadratic irrationality. $\square$

This means that $j(\alpha)$ is transcendental for every algebraic α in the upper half-plane which is not quadratic. On the other hand, if $\mathbb{Q}(\alpha)$ is imaginary quadratic, that is α is a CM point, one can show that $j(\alpha)$ is indeed an algebraic number. This is really a chapter in class field theory. We give a brief indication of why $j(\alpha)$ is algebraic in this case.

Recall that the ring of endomorphisms $\text{End}(E)$ of an elliptic curve E is either $\mathbb{Z}$ or an order in an imaginary quadratic field. In the latter case, we say the curve has *complex multiplication*.

Let $\mathcal{O}$ be any order in the ring of integers of $k = \mathbb{Q}(\alpha)$. As we mentioned in the previous chapter, the set of equivalence classes of invertible fractional ideals of $\mathcal{O}$ forms a multiplicative abelian group called the *Picard group* of $\mathcal{O}$ and there is a one-to-one correspondence between isomorphism classes of elliptic curves whose endomorphism ring is isomorphic to $\mathcal{O}$ and ideal classes of the Picard group of $\mathcal{O}$. This correspondence is given by taking an ideal $\mathfrak{a}$ of a given class and considering the elliptic curve $\mathbb{C}/\mathfrak{a}$. It is a standard theorem of algebraic number theory that this group is finite.

Now consider the lattice $L = \mathbb{Z}\alpha + \mathbb{Z}$ and let E_α be an elliptic curve whose period lattice is L. Thus $j(E_\alpha) = j(\alpha)$. Let $\text{End}(E_\alpha) = \mathcal{O}$ where $\mathcal{O}$ is an order in the ring of integers of k. Now for any automorphism σ of $\mathbb{C}$, let E_α^σ denote the curve obtained by applying σ to g_2, g_3. Clearly $j(E_\alpha^\sigma) = j(E_\alpha)^\sigma$ and $\text{End}(E_\alpha^\sigma) \simeq \text{End}(E_\alpha) = \mathcal{O}$. But there are only finitely many isomorphism classes of elliptic curves with a fixed endomorphism ring. Therefore the set of values $j(\alpha)^\sigma$ as σ ranges over automorphisms of $\mathbb{C}$ is a finite set and thus $j(\alpha)$ is necessarily algebraic. For a more detailed account, see the books by Lang [77] and Silverman [115].

In fact, if α is an imaginary quadratic irrational, then $j(\alpha)$ is an algebraic integer. Furthermore, the degree of $j(\alpha)$ is equal to the class number of $\mathbb{Q}(\alpha)$ (see [112] or [115]).

Let us consider the following interesting example by letting

$$\alpha = \frac{1 + \sqrt{-163}}{2}.$$

The field $\mathbb{Q}(\sqrt{-163})$ has class number one. In fact it is the "largest" imaginary quadratic field with class number one. More precisely, there exists no squarefree integer $d > 163$ such that $\mathbb{Q}(\sqrt{-d})$ has class number one.

Now for any z in the upper half-plane, the j-function has the following expansion

$$j(z) = \frac{1}{q} + 744 + 196884q + \cdots$$

where $q = e^{2\pi i z}$. In the case $z = \alpha$, we have

$$j(\alpha) = -e^{\pi\sqrt{163}} + 744 - 196884 e^{-\pi\sqrt{163}} + \cdots$$

Now $j(\alpha)$ must be an ordinary integer as $\mathbb{Q}(\sqrt{-163})$ has class number one. Consequently, we have the following curious expression

$$e^{\pi\sqrt{163}} = 262537412640768743.99999999999925\ldots$$

$$= (640320)^3 + 744 + O\left(e^{-\pi\sqrt{163}}\right)$$

and that $j(\alpha) = -(640320)^3$. Note that $e^{\pi\sqrt{163}}$ is a transcendental number by the Gelfond–Schneider theorem.

Exercises

1. Show that
$$e^{\pi\sqrt{67}} = 147197952743.9999999999\ldots$$
   accurate to ten decimal places.

2. Deduce that
$$j\left(\frac{1 + \sqrt{-67}}{2}\right) = -147197952000.$$

3. Show that $j(i) = 1728$.

4. Show that $j((1 + \sqrt{-3})/2) = 0$.

5. Let L be a lattice and $\wp$ be the associated Weierstrass function. Show that for any complex number α, $\alpha L \subset L$ if and only if $\wp(\alpha z)$ is a rational function in $\wp$.

6. Show that the group of field automorphisms of $\mathbb{C}$ is uncountable. What about the automorphisms of $\mathbb{R}$?

Chapter 16

More Elliptic Integrals

We will look at two explicit consequences of Schneider's theorem on the transcendence of periods of elliptic curves defined over the algebraic numbers.

Let us look at the curve

$$y^2 = 4x^3 - 4.$$

One of the periods is

$$\int_1^\infty \frac{dx}{\sqrt{x^3 - 1}}.$$

This can be related to the classical beta function as follows. Let us first put $x = 1/t$ to transform the integral to

$$\int_0^1 t^{-1/2}(1 - t^3)^{-1/2} dt.$$

Putting $t^3 = u$ changes it to

$$\frac{1}{3}\int_0^1 u^{-5/6}(1 - u)^{-1/2} du = \frac{1}{3}B(1/6, 1/2),$$

where

$$B(a, b) = \int_0^1 u^{a-1}(1 - u)^{b-1} du, \quad \Re(a), \Re(b) > 0.$$

Using the following formula for the beta function

$$B(a, b) = \frac{\Gamma(a)\Gamma(b)}{\Gamma(a + b)},$$

M.R. Murty and P. Rath, *Transcendental Numbers*, DOI 10.1007/978-1-4939-0832-5_16,   79
© Springer Science+Business Media New York 2014

one can show that the period is equal to

$$\frac{\Gamma(1/3)^3}{2^{4/3}\pi}.$$

The formula for the beta function is easily derived (see Chap. 18) by putting $u = \cos^2\theta$ which transforms the integral into

$$2\int_0^{\pi/2} \cos^{2a-1}\theta \sin^{2b-1}\theta d\theta.$$

Thus by Schneider's theorem, the number

$$\frac{\Gamma(1/3)^3}{\pi}$$

is transcendental.

Another curve to consider is

$$y^2 = 4x^3 - 4x.$$

One of the periods (say ω) is

$$\int_1^\infty \frac{dx}{\sqrt{x^3 - x}}.$$

By what we have proved, this integral is transcendental. Similarly as above, we find that this is

$$\frac{1}{2}\int_0^1 u^{-3/4}(1-u)^{-1/2} = \frac{1}{2}B(1/4,1/2).$$

By the previous identity involving the beta function, we have

$$\omega = \frac{\Gamma(1/4)^2}{2\sqrt{2\pi}}.$$

For the above curve

$$y^2 = 4x^3 - 4x,$$

we have $g_3 = 0$. Since $g_3(i) = i^6 g_3(i) = 0$, its lattice L is given by

$$L = \mathbb{Z}(i\omega) + \mathbb{Z}\omega.$$

Clearly this has complex multiplication by $\mathbb{Z}[i]$. Furthermore, this curve corresponds to the point $z = i$ in the standard fundamental domain and has j-invariant equal to 1728. We therefore deduce that

$$\sum_{(m,n)\neq(0,0)} \frac{1}{(mi+n)^4} = \frac{1}{15}\frac{\Gamma(1/4)^8}{2^6\pi^2}$$

as this is simply the corresponding Eisenstein series evaluated at i.

These calculations can be generalised for CM elliptic curves. Indeed, if $\mathcal{O}$ is an order in an imaginary quadratic field K and E is an elliptic curve with CM by $\mathcal{O}$, then the corresponding lattice L determines a vector space $L \otimes \mathbb{Q}$. This is invariant under the action of K. Therefore $L \otimes \mathbb{Q} = K\omega$ for some $\omega \in \mathbb{C}^{\times}$ defined up to elements of $K^{\times}$. In particular, if $\mathcal{O} = \mathcal{O}_K$ is the full ring of integers of K, ω is given by the Chowla–Selberg formula:

$$\omega = \alpha\sqrt{\pi} \prod_{0<a<d,(a,d)=1} \Gamma(a/d)^{w\chi(a)/4h}$$

where α is an algebraic number, w is the number of roots of unity in K, $-d$ is the discriminant of K, χ is the quadratic character mod d determined by K and h is the class number of K. We shall come back to the Chowla–Selberg formula in Chap. 26.

In the special case of $y^2 = 4x^3 - 4x$, the formula gives

$$\alpha\sqrt{\pi}\Gamma(1/4)\Gamma(3/4)^{-1},$$

which is in agreement with our earlier formula once we apply the usual functional equations of the Γ-function to it.

Exercises

1. Show that the beta function $B(a, b)$ can be given in terms of the Γ-function as $\Gamma(a)\Gamma(b)/\Gamma(a + b)$ for $\Re(a), \Re(b) > 0$.

2. Define the complete elliptic integral of the first kind by

$$K(k) = \int_0^{\pi/2} \frac{d\theta}{\sqrt{1 - k^2 \sin^2 \theta}}.$$

Show that $K(1/\sqrt{2}) = \Gamma(1/4)^2/4\sqrt{\pi}$.

3. The complete elliptic integral of the second kind is given by

$$E(k) = \int_0^{\pi/2} \sqrt{1 - k^2 \sin^2 \theta}d\theta.$$

Show that

$$E(1/\sqrt{2}) = \pi^{3/2}\Gamma(1/4)^{-2} + \Gamma(1/4)^2/8\sqrt{\pi}.$$

4. Show that

$$\sum_{(m,n)\neq(0,0)} \frac{1}{(m\rho + n)^6} = \frac{\Gamma(1/3)^{18}}{5.7.2^8\pi^6},$$

where $\rho = e^{2\pi i/3}$.

Chapter 17

Transcendental Values of Eisenstein Series

In this chapter, we will apply the Schneider–Lang theorem to study the transcendental values of the Eisenstein series introduced in earlier chapters.

Theorem 17.1 *Let $\wp$ be a Weierstrass $\wp$-function with algebraic invariants g_2, g_3 and z_0 a complex number which is not a pole of $\wp$. Then at least one of the numbers $e^{z_0}, \wp(z_0)$ is transcendental.*

Proof. Suppose not. Let K be the field $\mathbb{Q}(g_2, g_3, e^{z_0}, \wp(z_0), \wp'(z_0))$. We apply the Schneider–Lang theorem to the ring generated by $K[f_1, f_2, f_3]$ where $f_1(z) = e^z$, $f_2(z) = \wp(z)$ and $f_3(z) = \wp'(z)$. We need to show that f_1, f_2 are algebraically independent, but this is easily done (see for instance, Exercise 1). By Schneider–Lang, there only finitely many values at which these functions can simultaneously take values in K. However, since e^{z_0} and $\wp(z_0)$ are in K, so are e^{nz_0} and $\wp(nz_0)$ for infinitely many $n \in \mathbb{N}$. This completes the proof. $\square$

We remark that if e^z is replaced by $e^{\beta z}$ with β algebraic, then a suitable modification of the proof leads to:

Theorem 17.2 *Let $\wp$ be as above with algebraic invariants g_2, g_3. Let $\beta \neq 0$ be algebraic and z_0 a complex number which is not a pole of $\wp(z)$. Then, at least one of $e^{\beta z_0}, \wp(z_0)$ is transcendental.*

Corollary 17.3 *At least one of*

$$g_2, g_3, \beta, \wp(\alpha), e^{\beta \alpha}$$

is transcendental.

M.R. Murty and P. Rath, *Transcendental Numbers*, DOI 10.1007/978-1-4939-0832-5_17,   83
© Springer Science+Business Media New York 2014

In the special case when g_2, g_3 are algebraic and $\wp(\alpha)$ and e^γ are algebraic with $\gamma \neq 0$, γ/α is transcendental. If not, we may apply the corollary with $\beta = \gamma/\alpha$ and derive a contradiction. In particular, we deduce that α/π is transcendental for any algebraic point α of $\wp$ (i.e. $\alpha \in \mathbb{C}$ such that $\wp(\alpha)$ is algebraic). Putting $\alpha = \omega/2$ where ω is a fundamental period, we derive the transcendence of ω/π. We record this as:

Corollary 17.4 *If α is an algebraic point of $\wp(z)$ and $\beta \neq 0$ is an algebraic number, then $e^{\beta\alpha}$ is transcendental. In particular, α/π is transcendental.*

Bertrand [17] observed that this result can be used to derive results about transcendental values of classical Eisenstein series. These were introduced in an earlier chapter. But we normalise these as follows. For $z \in \mathbb{H}$ and $q = e^{2\pi i z}$ (thus $0 < |q| < 1$), let us define the normalised Eisenstein series as

$$E_{2k}(q) = \frac{G_{2k}(z)}{2\zeta(2k)}$$

and hence by Theorem 14.4,

$$E_{2k}(q) = 1 - \frac{4k}{B_{2k}} \sum_{n=1}^{\infty} \sigma_{2k-1}(n)q^n.$$

Then we have

Theorem 17.5 (D. Bertrand) *For all complex numbers q with $0 < |q| < 1$, at least one of the numbers $E_4(q), E_6(q)$ is transcendental.*

Proof. Let $z \in \mathbb{C}$ with $\Re(z) < 0$ such that $q = e^z$. Consider the lattice L spanned by $2\pi i$ and z. This is a rank 2 lattice since $\Re(z) \neq 0$. The corresponding Weierstrass $\wp$-function has g_2 and g_3 given by rational multiples of

$$\sum_{(m,n)\neq(0,0)} (mz + 2\pi in)^{-2k}$$

for $k = 2, 3$. By Theorem 14.2, we see that g_2, g_3 are rational multiples of $E_4(q), E_6(q)$, respectively, where $q = e^z$. Observe that $z = i\pi$ is an algebraic point of $\wp$. Since $e^{i\pi} = -1$ is algebraic, this contradicts Theorem 17.1. This completes the proof. $\square$

We now describe some recent work of Nesterenko that generalises the theorem of Bertrand and as a consequence proves the algebraic independence of π and e^π.

With Ramanujan, we introduce the Eisenstein series

$$E_2(q) = 1 - 24 \sum_{n=1}^{\infty} \sigma_1(n)q^n.$$

Nesterenko proved:

Theorem 17.6 ([84]) *For each $q \in \mathbb{C}$, $0 < |q| < 1$, at least three of the numbers*

$$q, E_2(q), E_4(q), E_6(q)$$

are algebraically independent over $\mathbb{Q}$.

An immediate consequence is the following:

Corollary 17.7 *If q is an algebraic number with $0 < |q| < 1$, then $E_2(q)$, $E_4(q)$, $E_6(q)$ are algebraically independent over $\mathbb{Q}$. In particular, each of these numbers is transcendental.*

Another corollary is the following result originally conjectured by Mahler (see [81]) and first proved by Barré-Sirieix et al. [15] in 1995.

Corollary 17.8 *For any $\tau \in \mathbb{H}$, at least one of the two numbers $e^{2\pi i \tau}$ and $j(\tau)$ is transcendental.*

This follows from the identity

$$j(\tau) = 1728 \; \frac{E_4(q)^3}{E_4(q)^3 - E_6(q)^2} \;, \quad q = e^{2\pi i \tau}$$

which can be easily derived from the definitions of E_4 and E_6.

The proof of Barré-Sirieix et al. is based on modular arguments, different from those developed by Nesterenko. In this set-up, there is a general conjecture by Manin (see [82]) which states that for any algebraic number α different from 0 and 1 and any τ in the upper half-plane $\mathbb{H}$, at least one of the two numbers α^τ and $j(\tau)$ is transcendental. Here, $\alpha^\tau = e^{\tau \log \alpha}$ with any fixed choice of a branch of logarithm. This conjecture is open.

Another important consequence of Nesterenko's result is:

Corollary 17.9 *Let $\wp(z)$ be a Weierstrass $\wp$-function with algebraic invariants g_2, g_3. Let ω_1, ω_2 be its fundamental periods with $\omega_1/\omega_2 \in \mathbb{H}$. Let η_1, η_2 be the corresponding quasi-periods. Then,*

$$e^{2\pi i(\omega_1/\omega_2)}, \omega_2/\pi, \eta_2/\pi$$

are algebraically independent over $\mathbb{Q}$.

To deduce the corollary from Theorem 17.6, we use the fact that for $q = e^{2\pi i(\omega_1/\omega_2)}$, we have

$$E_2(q) = 3\frac{\omega_2}{\pi}\frac{\eta_2}{\pi}, \quad E_4(q) = \frac{3}{4}\left(\frac{\omega_2}{\pi}\right)^4 g_2, \quad E_6(q) = \frac{27}{8}\left(\frac{\omega_2}{\pi}\right)^6 g_3.$$

The last two are clear from our previous analysis. The first requires proof and this is somewhat delicate since E_2 is not a modular form (see [77], for instance).   These formulas imply that $E_2(q), E_4(q), E_6(q)$ are algebraic over the field $\mathbb{Q}(\omega_2/\pi, \eta_2/\pi)$. But by Theorem 17.6, the field generated by $q, E_2(q), E_4(q), E_6(q)$ has transcendence degree 3. The corollary now follows. We add that the algebraic independence of the two numbers ω_2/π and η_2/π was first established by Chudnovsky.

An interesting situation arises in the complex multiplication case. Let us first prove the following lemma proved by Masser [83].

Lemma 17.10 *Let $\wp(z)$ be a Weierstrass $\wp$-function with algebraic invariants g_2, g_3 and complex multiplication by an order in the imaginary quadratic field k. Let ω_1, ω_2 and η_1, η_2 be certain fundamental periods and quasi periods, respectively. Then ω_1 and η_1 are algebraic over the field $\mathbb{Q}(\omega_2, \eta_2)$.*

Proof. Let $K = k(g_2, g_3)$. Since $\tau = \omega_1/\omega_2 \in k$ and lies in $\mathbb{H}$, it satisfies an equation

$$a\tau^2 + b\tau + c = 0$$

with co-prime integers a, b, c and $a \neq 0$. Let

$$c\eta_2 - a\tau\eta_1 = \alpha\omega_1$$

for some α in $\mathbb{C}$. We will show that $\alpha \in K$ and this will prove the assertion. Let f be the function defined as

$$f(z) = -c\,\zeta(az) + a\tau\zeta(a\tau z) + a\tau\alpha z.$$

Then

$$f(z + \omega_2) - f(z) = -ca\eta_2 + a^2\tau\eta_1 + a\tau\alpha\omega_2 = 0.$$

Further since

$$a\tau\omega_1 = -b\omega_1 - c\omega_2,$$

a similar calculation shows that $f(z + \omega_1) = f(z)$. Thus, f is a doubly periodic function with respect to the lattice L of $\wp$ and hence is a rational function in $\wp(z)$ and $\wp'(z)$. Now for any embedding σ of $K(\alpha)$ in $\mathbb{C}$ fixing K, we can construct a new function f^σ by acting σ on the Laurent expansion of f around the origin. This again is a rational function in $\wp(z)$ and $\wp'(z)$ as σ fixes $\wp$ and $\wp'$. Thus

$$f(z) - f^\sigma(z) = a\tau z(\alpha - \sigma(\alpha))$$

is also an elliptic function and hence $\alpha = \sigma(\alpha)$. Since σ is arbitrary, we see that $\alpha \in K$. $\square$

From this and using the Legendre relation

$$\omega_1\eta_2 - \omega_2\eta_1 = 2\pi i,$$

we deduce immediately:

Corollary 17.11 *Let $\wp(z)$ be a Weierstrass $\wp$-function with algebraic invariants g_2, g_3 and with complex multiplication by an order of the imaginary quadratic field K. Let ω be a non-zero period and η the corresponding quasi-period. Then for any $\tau \in K$ with $\Im(\tau) \neq 0$, each of these sets*

$$\{\pi, \omega, e^{2\pi i \tau}\} \quad \text{and} \quad \{\omega, \eta, e^{2\pi i \tau}\}$$

is algebraically independent over $\mathbb{Q}$.

Applying this corollary to the two elliptic curves

$$y^2 = 4x^3 - 4x$$

and

$$y^2 = 4x^3 - 4$$

considered earlier leads us to:

Corollary 17.12 *Each of the sets*

$$\{\pi, e^\pi, \Gamma(1/4)\}, \qquad \{\pi, e^{\pi\sqrt{3}}, \Gamma(1/3)\}$$

is algebraically independent over $\mathbb{Q}$. In particular, π and e^π are algebraically independent. The same holds for $\pi, \Gamma(1/3)$ and for $\pi, \Gamma(1/4)$.

We reiterate that the algebraic independence of $\{\pi, \Gamma(1/3)\}$ as well as that of $\{\pi, \Gamma(1/4)\}$ was first established by Chudnovsky. We note that this is the only known way of deducing the irrationality of $\Gamma(1/3)$ and $\Gamma(1/4)$.

By the theory of complex multiplication, we know that for any squarefree natural number D, there is an elliptic curve with algebraic invariants and with complex multiplication by an order in $\mathbb{Q}(\sqrt{-D})$. Thus we deduce:

Corollary 17.13 *For any positive integer D, the numbers*

$$\pi \quad \text{and} \quad e^{\pi\sqrt{D}}$$

are algebraically independent over $\mathbb{Q}$.

In a later chapter, we shall apply Nesterenko's result to study the values taken by modular forms defined over number fields.

Exercises

1. Show that e^z and $\sigma(z)$ have different orders.

2. If β is algebraic and $\wp$ is a Weierstrass $\wp$-function with algebraic invariants, show that $\wp(2\pi i\beta)$ is transcendental.

3. Prove that e^z and the Weierstrass ζ-function are algebraically independent.

4. If α is an algebraic point of the Weierstrass ζ function and β is an algebraic number, show that $e^{\beta\alpha}$ is transcendental.

5. If $\wp$ has algebraic invariants g_2, g_3, show that η/π is transcendental, where η is a non-zero quasi-period of ζ.

6. Let $y \neq 0$ be a real number.

   (a) Prove that $|\Gamma(iy)|^2 = \frac{\pi}{y\sinh(\pi y)}$.

   (b) For y rational, show that $\Gamma(iy)$ is transcendental.

   (c) For $D > 0$ squarefree and any non-zero y in $\mathbb{Q}(\sqrt{D})$, show that $\Gamma(iy)$ is transcendental.

7. For any non-zero $y \in \mathbb{Q}$ and $n \in \mathbb{N}$, show that $\Gamma(n+iy)$ is transcendental.

8. For $D > 0$ squarefree and non-zero $y \in \mathbb{Q}(\sqrt{D})$, show that $\Gamma(n+iy)$ is transcendental.

9. Let $y \in \mathbb{R}$.

   (a) Prove that $|\Gamma(\frac{1}{2}+iy)|^2 = \frac{\pi}{\cosh(\pi y)}$.

   (b) For $y \in \mathbb{Q}$, show that $\Gamma(\frac{1}{2}+iy)$ is transcendental.

   (c) For $D > 0$ squarefree and $y \in \mathbb{Q}(\sqrt{D})$, show that $\Gamma(\frac{1}{2}+iy)$ is transcendental.

Chapter 18

Elliptic Integrals and Hypergeometric Series

We have already discussed briefly the problem of inversion for the Weierstrass $\wp$-function. In this way, we were able to recover the transcendental nature of the periods whenever the invariants g_2, g_3 were algebraic. We now look at the calculation a bit more closely. Before we begin, it may be instructive to look at a familiar example. Clearly, we have

$$b = \int_0^{\sin b} \frac{dy}{\sqrt{1-y^2}}.$$

But how should we view this equation? Since $\sin b$ is periodic with period 2π, we can only view this as an equation modulo 2π. If $\sin b$ is algebraic, then, we know as a consequence of the Hermite–Lindemann theorem that b is transcendental. In this way, we deduce that the integral

$$\int_0^{\alpha} \frac{dy}{\sqrt{1-y^2}}$$

is transcendental whenever α is a non-zero algebraic number in the interval $[-1, 1]$.

A similar result can be obtained for incomplete elliptic integrals. Recall that we have written our elliptic curve as

$$y^2 = 4(x - e_1)(x - e_2)(x - e_3).$$

M.R. Murty and P. Rath, *Transcendental Numbers*, DOI 10.1007/978-1-4939-0832-5_18,   89
© Springer Science+Business Media New York 2014

To reiterate, it is appropriate to consider the extended complex plane with the point at infinity added and to look at paths in this region (see [67]). With this in mind, as before we obtain

$$z = \int_\infty^{\wp(z)} \frac{dx}{\sqrt{4x^3 - g_2 x - g_3}}$$

which is again to be interpreted as up to periods.

We will need the following fact: For s with $\Re(s) > 0$, we have

$$\Gamma(s) = \int_0^\infty e^{-x} x^s \frac{dx}{x},$$

and putting $x = t^2$ gives

$$\Gamma(s) = 2 \int_0^\infty t^{2s} e^{-t^2} \frac{dt}{t}.$$

We will use this to show the following for $a, b > 0$.

$$2 \int_0^{\pi/2} \cos^{2a-1} \theta \sin^{2b-1} \theta d\theta = \frac{\Gamma(a)\Gamma(b)}{\Gamma(a+b)}.$$

Indeed, we calculate

$$\Gamma(a)\Gamma(b) = 4 \int_0^\infty \int_0^\infty x^{2a-1} y^{2b-1} \exp(-x^2 - y^2) dx dy,$$

and switching to polar co-ordinates, we get that this is

$$2 \int_0^\infty r^{2a+2b} e^{-r^2} \frac{dr}{r} 2 \int_0^{\pi/2} \cos^{2a-1} \theta \sin^{2b-1} \theta d\theta.$$

The special case $a = b = 1/2$ shows that $\Gamma(1/2) = \sqrt{\pi}$.

We can consider our elliptic curve in *Legendre normal form*, that is, of the form

$$E_\lambda : \quad y^2 = x(x-1)(x-\lambda)$$

where $\lambda \in \mathbb{C}\backslash\{0,1\}$. In fact, this change of variable works over any field of characteristic not equal to 2. The j-invariant of E_λ is easily computed (see exercise below):

$$j(E_\lambda) = \frac{2^8(\lambda^2 - \lambda + 1)^3}{\lambda^2(\lambda-1)^2}.$$

If we change x to λx and y to $\lambda^{3/2} y$, then the curve is isomorphic over $\mathbb{C}$ to

$$y^2 = x(x-1)(x-1/\lambda).$$

Notice that if λ is algebraic, then the change of variables is again algebraic. Thus, we may suppose (without any loss of generality) that there is a model for E with $|\lambda| < 1$. We may express the periods (see [67], for instance) as

$$\omega_1(\lambda) = \int_{-\infty}^{0} \frac{dx}{\sqrt{x(x-1)(x-\lambda)}}$$

and

$$\omega_2(\lambda) = \int_{1}^{\infty} \frac{dx}{\sqrt{x(x-1)(x-\lambda)}}.$$

We now recall the hypergeometric series: for $a, b \in \mathbb{C}$ and $c \in \mathbb{C}\backslash\mathbb{N}$, we define

$$F(a, b, c; z) = \sum_{n=0}^{\infty} \frac{(a)_n (b)_n}{n!(c)_n} z^n,$$

where

$$(a)_n = a(a+1)\cdots(a+n-1), \quad (a)_0 = 1.$$

A straightforward application of the ratio test shows that this series converges absolutely for $|z| < 1$ (see exercises below). Thus, it represents an analytic function in this disc.

It is clear that $F(a, b, c; z) = F(b, a, c; z)$ and that

$$F(a, b, b; z) = (1-z)^{-a}.$$

It is also not hard to see that

$$F(a, a, 1; z) = \sum_{n=0}^{\infty} \binom{-a}{n}^2 z^n.$$

The hypergeometric series satisfies the following differential equation:

$$z(1-z)F'' + (c - (a+b+1)z)F' - abF = 0.$$

Theorem 18.1 *For a complex number λ with $|\lambda| < 1$,*

$$2\int_{0}^{\pi/2} (1 - \lambda\sin^2\theta)^{-1/2}d\theta = \pi F(1/2, 1/2, 1; \lambda).$$

Proof. We use the binomial theorem to expand the integrand as

$$(1 - \lambda\sin^2\theta)^{-1/2} = \sum_{n=0}^{\infty} \binom{-1/2}{n}(-\lambda)^n \sin^{2n}\theta.$$

Integrating this term by term and using Exercise 1, we get the result. $\square$

Let us again consider the integral

$$\omega_2(\lambda) = \int_{1}^{\infty} \frac{dx}{\sqrt{x(x-1)(x-\lambda)}}.$$

Putting $x = 1/t, t = s^2, s = \sin\theta$ in succession, transforms the integral into

$$\int_0^1 \frac{dt}{\sqrt{t(1-t)(1-\lambda t)}} = 2\int_0^1 \frac{ds}{\sqrt{(1-s^2)(1-\lambda s^2)}}$$

$$= 2\int_0^{\pi/2} \frac{d\theta}{\sqrt{1-\lambda\sin^2\theta}} = \pi F(1/2, 1/2, 1; \lambda)$$

A similar calculation for $\omega_1(\lambda)$ shows that the other period is

$$\omega_1(\lambda) = \int_{-\infty}^0 \frac{dx}{\sqrt{x(x-1)(x-\lambda)}} = i\pi F(1/2, 1/2, 1, 1-\lambda).$$

In the case $\lambda = 1/2$, observe that $\omega_1(1/2) = i\omega_2(1/2)$ and hence the quotient of these two periods is equal to i. An immediate consequence of Schneider's theorem is

Theorem 18.2 *For algebraic λ with $|\lambda| < 1$, both the numbers*

$$\pi F(1/2, 1/2, 1, \lambda) \quad \text{and} \quad F(1/2, 1/2, 1, \lambda)$$

are transcendental.

Proof. The first number is transcendental since it is a period of an elliptic curve defined over $\overline{\mathbb{Q}}$. The second number is transcendental since it is this period divided by π. $\square$

Recall that in calculating the circumference of an ellipse with major axis and minor axis of lengths a and b, respectively, we show that it is given by

$$4\int_0^{\pi/2} \sqrt{a^2\cos^2\theta + b^2\sin^2\theta}\, d\theta = 4\int_0^{\pi/2} \sqrt{a^2 - (a^2 - b^2)\sin^2\theta}\, d\theta,$$

as is easily seen by putting $\cos^2\theta = 1 - \sin^2\theta$. We can re-write this integral as

$$4a\int_0^{\pi/2} \sqrt{1-\lambda\sin^2\theta}\, d\theta,$$

where $\lambda = 1 - b^2/a^2$. We may expand the integral via the binomial theorem to get

$$4a\int_0^{\pi/2} \sum_{n=0}^{\infty} \binom{1/2}{n} (-1)^n \lambda^n \sin^{2n}\theta\, d\theta.$$

Using the result

$$2\int_0^{\pi/2} \cos^{2a-1}\theta\sin^{2b-1}\theta\, d\theta = \frac{\Gamma(a)\Gamma(b)}{\Gamma(a+b)},$$

we see that we may apply this with $a = 1/2, b = n + 1/2$ to get

$$2 \int_0^{\pi/2} \sin^{2n} \theta d\theta = \frac{\Gamma(1/2)\Gamma(n+1/2)}{\Gamma(n+1)}.$$

The last term can be re-written as

$$\frac{\Gamma(1/2)(n+1/2-1)(n+1/2-2)\cdots(n+1/2-n)\Gamma(1/2)}{n!} = \frac{\pi(1/2)_n}{n!}.$$

Putting this all together, we obtain:

Theorem 18.3 *The circumference of an ellipse with major and minor axes of lengths a and b, respectively, is*

$$2\pi a \sum_{n=0}^{\infty} \binom{1/2}{n} (-1)^n \frac{(1/2)_n}{n!} \lambda^n$$

where $\lambda = 1 - b^2/a^2$.

The series is in fact a hypergeometric series as is easily seen by noting that

$$(-1)^n \binom{1/2}{n} = \frac{(-1/2)_n}{n!}.$$

Thus, the circumference of the ellipse is

$$2\pi a F(-1/2, 1/2, 1; \lambda).$$

There has been some work in trying to determine for which arguments the general hypergeometric function takes transcendental values. In the case a, b, c are rational numbers, with $c \neq 0, -1, -2, \ldots$, a theorem of Wolfart states that if $F(a, b, c; z)$ is not algebraic over $\mathbb{C}(z)$ and its monodromy group is not an arithmetic hyperbolic triangle group, then there are only finitely many values of $z \in \overline{\mathbb{Q}}$ for which $F(a, b, c; z)$ is algebraic.

Exercises

1. Show that for any natural number n,

$$2 \int_0^{\pi/2} \sin^{2n} \theta d\theta = \pi(-1)^n \binom{-\frac{1}{2}}{n}.$$

2. Show that the area of the ellipse given by

$$\frac{x^2}{a^2} + \frac{y^2}{b^2} = 1$$

is πab.

3. Show that the j-invariant of E_λ is given by

$$j(E_\lambda) = \frac{2^8(\lambda^2 - \lambda + 1)^3}{\lambda^2(\lambda - 1)^2}.$$

4. Determine the radius of convergence for the hypergeometric series.

5. Show that $F(a, b, b; z) = (1 - z)^{-a}$.

6.

$$F(a, a, 1; z) = \sum_{n=0}^{\infty} \binom{-a}{n}^2 z^n.$$

Chapter 19

Baker's Theorem

In this chapter, we will discuss the following theorem due to Baker.

Theorem 19.1 ([8]) *If $\alpha_1, \ldots, \alpha_m$ are non-zero algebraic numbers such that $\log \alpha_1, \ldots, \log \alpha_m$ are linearly independent over $\mathbb{Q}$, then*

$$1, \log \alpha_1, \ldots, \log \alpha_m$$

are linearly independent over $\overline{\mathbb{Q}}$.

Observe that the case $m = 1$ is a consequence of the Lindemann–Weierstrass theorem. The case $m = 2$ implies the Gelfond–Schneider theorem. In 1980, Bertrand and Masser [18] proved an elliptic analog of Baker's theorem. For a Weierstrass $\wp$-function with algebraic invariants g_2 and g_3 and field of endomorphisms k, the following set

$$\mathcal{L}_E = \{\alpha \in \mathbb{C} \ : \ \wp(\alpha) \in \overline{\mathbb{Q}} \cup \{\infty\}\}$$

is referred to as the set of elliptic logarithms of algebraic points on E. Here E is the associated elliptic curve. Let L be the lattice of periods. The k-linear space $\mathcal{L}_E$ is the elliptic analog of the $\mathbb{Q}$-linear space of logarithms of non-zero algebraic numbers for the exponential case. Bertrand and Masser proved the following theorem.

Theorem 19.2 *Let $\wp$ be a Weierstrass function without complex multiplication and with algebraic invariants g_2, g_3. Let $u_1, \ldots, u_n$ be elements in $\mathcal{L}_E$ such that $u_1, \ldots, u_n$ are linearly independent over $\mathbb{Q}$. Then*

$$1, u_1, \ldots, u_n$$

are linearly independent over $\overline{\mathbb{Q}}$.

M.R. Murty and P. Rath, *Transcendental Numbers*, DOI 10.1007/978-1-4939-0832-5_19,   95
© Springer Science+Business Media New York 2014

The analogous theorem for the CM case was earlier established by Masser [83] and can also be recovered following the techniques employed in the proof of the above.

As noted by Bertrand and Masser themselves, their method gives an alternate proof of Baker's theorem. In this chapter, we present their proof of Baker's theorem. We also recommend the book by Waldschmidt [125] which includes several different proofs of Baker's theorem.

We begin by noting that Baker's theorem is a consequence of the following theorem.

Theorem 19.3 (Main Theorem) *Let K be a number field of degree d over $\mathbb{Q}$. Let $\beta_1, \ldots, \beta_d$ be a basis for K over $\mathbb{Q}$. Let $\alpha_1, \ldots, \alpha_d$ be non-zero algebraic numbers. Then*

$$\beta_1 \log \alpha_1 + \cdots + \beta_d \log \alpha_d \in \overline{\mathbb{Q}}$$

if and only if

$$\log \alpha_1 = \cdots = \log \alpha_d = 0.$$

Let us first see how the above theorem implies Baker's theorem. Let $\alpha_1, \ldots, \alpha_d$ be non-zero algebraic numbers such that $\log \alpha_1, \ldots, \log \alpha_d$ are linearly independent over $\mathbb{Q}$. Now suppose that

$$\beta_1 \log \alpha_1 + \cdots + \beta_d \log \alpha_d = \gamma$$

where $\beta_1, \ldots, \beta_d$ and γ are algebraic numbers. Consider the number field $K = \mathbb{Q}(\beta_1, \ldots, \beta_d)$ and let $x_1, \ldots, x_n$ be a $\mathbb{Q}$-basis for K. Let

$$\beta_i = \sum_{j=1}^{n} y_{ij} x_j$$

for $1 \leq i \leq d$, where y_{ij} are rational numbers. Thus, we have

$$\sum_{j=1}^{n} A_j x_j = \gamma \quad \text{where} \quad A_j = \sum_{i=1}^{d} y_{ij} \log \alpha_i, \quad 1 \leq j \leq n.$$

But the A_j's are logarithms of algebraic numbers as y_{ij}'s are rational numbers. By the above theorem, each A_j is necessarily equal to zero. But since $\log \alpha_1, \ldots, \log \alpha_d$ are linearly independent over $\mathbb{Q}$, this implies that $y_{ij} = 0$ for all i and j. Thus $\beta_i = 0$ for $1 \leq i \leq d$. Hence the above theorem implies Baker's theorem.

Let us now begin the proof of the main theorem. The crucial ingredient in the proof is the following multi-variable generalisation of the Schneider–Lang theorem we proved earlier. This was proved by Lang [78]. An entire function $F(\overline{z})$ in r variables, with $\overline{z} = (z_1, \ldots, z_r) \in \mathbb{C}^r$, is said to be of finite order of growth if

$$\limsup_{R \to \infty} \frac{\log \log |F|_R}{\log R} < \infty,$$

where $|F|_R$ is the supremum of $|F(\overline{z})|$ on the closed disc $|\overline{z}| \leq R$.

Theorem 19.4 ([78]) *For integers $N > r \geq 1$, let $f_1, \ldots, f_N$ be entire functions on $\mathbb{C}^r$ with finite order of growth and of which at least $r + 1$ are algebraically independent. Let K be a number field such that the ring $K[f_1, \ldots, f_N]$ is mapped into itself by the partial derivatives $\frac{\partial}{\partial z_1}, \ldots, \frac{\partial}{\partial z_r}$. Then for any subgroup Γ of $\mathbb{C}^r$ which contains a basis of the complex space $\mathbb{C}^r$, not all the values*

$$f_k(\bar{z}), \qquad 1 \leq k \leq N, \ \bar{z} \in \Gamma$$

can lie in K.

The proof of the above theorem, though more involved, runs along similar lines as in the one-dimensional case. We refer to chapter IV of Lang's book [79] for the relevant details.

The theory of several complex variables constitutes an essential tool in the development of modern transcendence theory. We refer to the classic treatise of Gunning and Rossi [64] for the basic definitions and notions. However for our purposes, we only need to work with very special type of entire functions, namely functions of the form

$$f(\bar{z}) = e^{\bar{a}.\bar{z}} = e^{a_1 z_1 + \cdots + a_r z_r}$$

where $\bar{a} = (a_1, \ldots, a_r)$ is a fixed vector in $\mathbb{C}^r$. Clearly, these functions have finite order of growth.

We deduce the following two corollaries of the above theorem. As before, for $\bar{x} = (x_1, \ldots, x_d), \ \bar{y} = (y_1, \ldots, y_d) \in \mathbb{C}^d$, we have the following notation

$$\bar{x}.\bar{y} = x_1 y_1 + \cdots + x_d y_d.$$

Corollary 19.5 *Let N, d with $N > d$ be positive integers and $\bar{x}_1, \ldots, \bar{x}_N$ be elements in $\overline{\mathbb{Q}}^d$ such that at least $d + 1$ of these vectors are linearly independent over $\mathbb{Q}$. Let $\bar{y}_1, \ldots, \bar{y}_M$ be elements in $\mathbb{C}^d$ containing a basis for $\mathbb{C}^d$. Then not all the MN numbers*

$$e^{\bar{x}_i . \bar{y}_j}$$

can be algebraic.

Proof. Consider the N functions

$$f_i(z_1, \ldots, z_d) = e^{\bar{x}_i.\bar{z}} = e^{x_{i1} z_1 + \cdots + x_{id} z_d}, \ 1 \leq i \leq N.$$

We note that at least $d + 1$ these functions are algebraically independent (see Exercise 1). Let Γ be the additive subgroup of $\mathbb{C}^d$ generated by the vectors $\bar{y}_1, \ldots, \bar{y}_M$. Suppose that the $e^{\bar{x}_i.\bar{y}_j}$ are all algebraic. Let K be the number field generated by the numbers $e^{\bar{x}_i.\bar{y}_j}$ and the coordinates of each of the vectors $\bar{x}_i = (x_{i1}, \ldots, x_{id})$. Then clearly all the hypotheses of Lang's theorem are satisfied. However, for all $1 \leq i \leq N$ and $\bar{z} \in \Gamma$, we have $f_i(\bar{z}) \in K$. This contradicts Lang's theorem. $\square$

Corollary 19.6 *Let $\overline{x}_1, \ldots, \overline{x}_d$ be d elements in $\overline{\mathbb{Q}}^d$ which are linearly independent over $\mathbb{Q}$. Let $\overline{y}_1, \ldots, \overline{y}_d$ be elements in $\mathbb{C}^d$ linearly independent over $\mathbb{C}$. For $1 \leq j \leq d$, let $\overline{y}_j = (y_{j1}, \ldots, y_{jd})$ and suppose that the kth entries of each of these d vectors, namely $y_{1k}, \ldots, y_{dk}$ are all algebraic. Then not all the d^2 numbers*

$$e^{\overline{x}_i \cdot \overline{y}_j}$$

can be algebraic.

Proof. As before, let

$$f_i(\overline{z}) = e^{\overline{x}_i \cdot \overline{z}}$$

for $1 \leq i \leq d$ and define $f_{d+1}(\overline{z}) = z_k$, the kth projection function. Let K be the number field generated by the numbers $e^{\overline{x}_i \cdot \overline{y}_j}$, the co-ordinates x_{ij} of the vectors $\overline{x}_i = (x_{i1}, \ldots, x_{id})$ and the kth coordinates y_{jk} of the vectors $\overline{y}_j$. Taking Γ to be the additive group generated by the vectors $\overline{y}_j$, we see that for all $1 \leq i \leq d+1$ and $\overline{z} \in \Gamma$, $f_i(\overline{z}) \in K$. This again contradicts Lang's theorem. $\square$

Let us now prove the main theorem. We have a number field K of degree d with $\beta_1, \ldots, \beta_d \in K$ constituting a basis. Further, $\alpha_1, \ldots, \alpha_d$ are non-zero algebraic numbers such that

$$\lambda = \beta_1 \log \alpha_1 + \cdots + \beta_d \log \alpha_d \in \overline{\mathbb{Q}}.$$

Our goal is to prove that

$$\log \alpha_1 = \cdots = \log \alpha_d = 0.$$

Let $\{\sigma_1, \ldots, \sigma_d\}$ be the embeddings of K in $\mathbb{C}$. We define the following complex numbers

$$\lambda_i = \sigma_i(\beta_1) \log \alpha_1 + \cdots + \sigma_i(\beta_d) \log \alpha_d$$

for $1 \leq i \leq d$. Consider the matrix M defined as

$$M = (\sigma_i(\beta_j))_{1 \leq i,j \leq d}.$$

Note that this matrix is non-singular (see Exercise 2). Thus if each of the λ_i is equal to zero, necessarily $\log \alpha_i$'s are all equal to zero. So we may assume that not all the λ_i's are equal to zero.

We first consider the case when none of the λ_i's are equal to zero, that is

$$\lambda_1 \ldots \lambda_d \neq 0.$$

We now construct vectors $\overline{x}_i$ and $\overline{y}_j$ as in Corollary 19.6. First for $1 \leq i \leq d$, let

$$\overline{x}_i = (\sigma_1(\beta_i), \ldots, \sigma_d(\beta_i)),$$

be the vector consisting of all the conjugates of β_i in $\mathbb{C}$. Non-singularity of the matrix M introduced above ensures that these d elements in $\overline{\mathbb{Q}}^d$ are linearly independent over $\mathbb{Q}$. Now we define the vectors $\overline{y}_j$ for $1 \leq j \leq d$ as follows:

$$\overline{y}_j = (\lambda_1 \sigma_1(\beta_j), \ldots, \lambda_d \sigma_d(\beta_j)).$$

Consider the matrix $(\lambda_i \sigma_i(\beta_j))_{1 \leq i,j \leq d}$. Its determinant is non-zero, being equal to $\det(M)\lambda_1 \ldots \lambda_d$. Thus the d vectors above are linearly independent over $\mathbb{C}$. If σ_k is the identity embedding of K, then the kth entries of these d vectors are given by $y_{jk} = \lambda \beta_j$ which are algebraic numbers for all $1 \leq j \leq d$. Thus we are in the situation to apply Corollary 19.6 which implies that not all the d^2 numbers $e^{\overline{x}_i \cdot \overline{y}_j}$ can be algebraic. Let us now explicitly evaluate the numbers $\overline{x}_i \cdot \overline{y}_j$. We have

$$\overline{x}_i \cdot \overline{y}_j = \sum_{l=1}^{d} \sigma_l(\beta_i) \lambda_l \sigma_l(\beta_j)$$

$$= \sum_{l=1}^{d} \sigma_l(\beta_i \beta_j) \lambda_l$$

$$= \sum_{l=1}^{d} \sum_{s=1}^{d} \sigma_l(\beta_i \beta_j) \sigma_l(\beta_s) \log \alpha_s$$

$$= \sum_{s=1}^{d} \left(\sum_{l=1}^{d} \sigma_l(\beta_i \beta_j \beta_s) \right) \log \alpha_s.$$

However, the number $A_s = \sum_{l=1}^{d} \sigma_l(\beta_i \beta_j \beta_s)$ is the trace of $\beta_i \beta_j \beta_s$ in K and hence rational. Thus,

$$e^{\overline{x}_i \cdot \overline{y}_j} = e^{A_1 \log \alpha_1 \cdots + A_d \log \alpha_d} \in \overline{\mathbb{Q}}.$$

This is a contradiction.

In the second case, suppose that some of the λ_i's are equal to zero. Without loss of generality, suppose that

$$\lambda_1 \neq 0, \ldots, \lambda_r \neq 0, \lambda_{r+1} = \cdots = \lambda_d = 0$$

with $1 \leq r < d$. In this case, we define the vectors $\overline{x}_i$ and $\overline{y}_j$ as follows. For $1 \leq i \leq d$, we define

$$\overline{x}_i = (\sigma_1(\beta_i), \ldots, \sigma_r(\beta_i)) \in \overline{\mathbb{Q}}^r$$

and for $1 \leq j \leq d$, we have

$$\overline{y}_j = (\lambda_1 \sigma_1(\beta_j), \ldots, \lambda_r \sigma_r(\beta_j)) \in \overline{\mathbb{C}}^r.$$

Note that the rank of the following $r \times d$ matrix

$$M = (\lambda_i \sigma_i(\beta_j))$$

where $1 \leq i \leq r$ and $1 \leq j \leq d$ is equal to r. Thus the d vectors $\overline{y}_j$ do contain a basis for $\mathbb{C}^r$. Again by Corollary 19.5, not all the d^2 numbers $e^{\overline{x}_i \cdot \overline{y}_j}$ can be algebraic. But an explicit evaluation of the numbers $\overline{x}_i.\overline{y}_j$ as above will show that

$$e^{\overline{x}_i \cdot \overline{y}_j} \in \overline{\mathbb{Q}}$$

for all i and j, hence a contradiction. This completes the proof of the main theorem and hence Baker's theorem.

Exercises

1. $\overline{x}_1, \ldots, \overline{x}_N$ be N elements in $\overline{\mathbb{Q}}^d$ which are linearly independent over $\mathbb{Q}$. Consider the N functions

$$f_i(z_1, \ldots, z_d) = e^{\overline{x}_i \cdot \overline{z}} = e^{x_{i1}z_1 + \cdots + x_{id}z_d}, \quad 1 \leq i \leq N.$$

   Show that these functions are algebraically independent.

2. Let G be an abelian group and $\sigma_1, \ldots, \sigma_d : G \to \mathbb{C}^\times$ be d distinct homomorphisms. Prove that these functions are linearly independent over $\mathbb{C}$. Hence conclude that the matrix M in the proof of the main theorem is invertible.

3. Let $\alpha_1, \alpha_2, \ldots, \alpha_n$ be positive algebraic numbers. If $c_0, c_1, \ldots, c_n$ are algebraic numbers with $c_0 \neq 0$, then show that

$$c_0 \pi + \sum_{j=1}^{n} c_j \log \alpha_j$$

   is a transcendental number and hence non-zero.

4. Show that

$$\int_0^1 \frac{dx}{1 + x^3}$$

   is transcendental. Can you generalise to rational functions with algebraic coefficients? (See [122].)

Chapter 20

Some Applications of Baker's Theorem

Let us first derive some important corollaries of Baker's theorem.

Corollary 20.1 *If $\alpha_1, \ldots, \alpha_m$ and $\beta_1, \ldots, \beta_m$ are algebraic with α_i's non-zero, then*

$$\beta_1 \log \alpha_1 + \cdots + \beta_m \log \alpha_m$$

is either zero or transcendental.

Proof. We proceed by induction on m. This clearly holds for $m = 1$ by the Lindemann–Weierstrass theorem. Now assume the validity of the corollary for $m < n$. We now proceed to prove it for $m = n$. Suppose not. Then

$$\beta_1 \log \alpha_1 + \cdots + \beta_n \log \alpha_n = \beta_0 \tag{20.1}$$

is algebraic and β_0 is non-zero. By Theorem 19.1, $\log \alpha_1, \ldots, \log \alpha_n$ must be linearly dependent over $\mathbb{Q}$. That is, there exist rational numbers $c_1, \ldots, c_n$, not all zero such that

$$c_1 \log \alpha_1 + \cdots + c_n \log \alpha_n = 0. \tag{20.2}$$

Say that $c_n \neq 0$, without any loss of generality. Using this relation, we can eliminate $\log \alpha_n$ from our original relation (20.1) and deduce a contradiction by induction. Indeed, multiplying (20.1) by c_n and relation (20.2) by β_n and subtracting, we get that

$$\beta_1' \log \alpha_1 + \cdots + \beta_{n-1}' \log \alpha_{n-1} = c_n \beta_0$$

which is not zero. We can now apply induction to deduce the corollary. $\square$

M.R. Murty and P. Rath, *Transcendental Numbers*, DOI 10.1007/978-1-4939-0832-5_20, 101
© Springer Science+Business Media New York 2014

Thus any algebraic linear combination of logarithms of algebraic numbers is either zero or transcendental. The next corollary represents a vast generalisation of the Gelfond–Schneider theorem.

Corollary 20.2 *If $\alpha_1, \ldots, \alpha_m$ and $\beta_0, \ldots, \beta_m$ are non-zero algebraic numbers, then*

$$e^{\beta_0} \alpha_1^{\beta_1} \cdots \alpha_m^{\beta_m}$$

is transcendental.

Proof. If the number were algebraic and equal to α_{n+1} say, then we get

$$\beta_1 \log \alpha_1 + \cdots + \beta_m \log \alpha_m - \log \alpha_{n+1} = -\beta_0 \neq 0.$$

This is a contradiction to the previous corollary. $\square$

Corollary 20.3 $\alpha_1^{\beta_1} \cdots \alpha_m^{\beta_m}$ *is transcendental for any algebraic numbers $\alpha_1, \ldots, \alpha_m$ other than 0 or 1 and any algebraic numbers $\beta_1, \ldots, \beta_m$ with 1, $\beta_1, \ldots, \beta_m$ linearly independent over the rationals.*

Proof. It suffices to show that for any algebraic numbers $\alpha_1, \ldots, \alpha_m$ other than 0 or 1 and any algebraic numbers $\beta_1, \ldots, \beta_m$ linearly independent over the rationals, we have

$$\beta_1 \log \alpha_1 + \cdots \beta_m \log \alpha_m \neq 0.$$

If we have this for every m, then we can apply this result with m replaced by $m + 1$ and $\beta_{m+1} = -1$ to derive a contradiction (since $-1, \beta_1, \ldots, \beta_m$ are linearly independent over $\mathbb{Q}$). We therefore proceed by induction on m which is clearly true for $m = 1$. Suppose we have proved it for $n < m$. If $\log \alpha_1, \ldots, \log \alpha_m$ are linearly independent over $\mathbb{Q}$, then the result follows from Theorem 19.1. So let us suppose otherwise. Then there are rational numbers $c_1, \ldots, c_m$ not all zero such that

$$c_1 \log \alpha_1 + \cdots + c_m \log \alpha_m = 0.$$

Without any loss of generality, let us suppose $c_m \neq 0$. We may use this relation to eliminate $\log \alpha_m$ to obtain

$$(c_m \beta_1 - c_1 \beta_m) \log \alpha_1 + \cdots + (c_m \beta_{m-1} - c_{m-1} \beta_m) \log \alpha_{m-1} = 0.$$

But the $m - 1$ numbers

$$c_m \beta_1 - c_1 \beta_m, \ldots, c_m \beta_{m-1} - c_{m-1} \beta_m$$

are linearly independent over $\mathbb{Q}$ for otherwise,

$$A_1(c_m \beta_1 - c_1 \beta_m) + \cdots + A_{m-1}(c_m \beta_{m-1} - c_{m-1} \beta_m) = 0$$

for some rational numbers $A_1, \ldots, A_{m-1}$ not all zero. But re-arranging this, we find

$$c_m(A_1\beta_1 + \cdots + A_{m-1}\beta_{m-1}) - (A_1c_1 + \cdots + A_{m-1}c_{m-1})\beta_m = 0.$$

Since $\beta_1, \ldots, \beta_m$ are linearly independent over $\mathbb{Q}$, we deduce that $A_1 = \cdots = A_{m-1} = 0$, a contradiction. This proves the corollary. $\square$

Corollary 20.4 $\pi + \log\alpha$ *is transcendental for any algebraic number* $\alpha \neq 0$. $e^{\alpha\pi+\beta}$ *is transcendental for any algebraic numbers* α, β *with* $\beta \neq 0$.

In 1966, Baker proved a quantitative version of his theorem. Such versions now fall under the general heading of effective lower bounds for linear forms in logarithms.

Theorem 20.5 ([8]) *Let* $\alpha_1, \ldots, \alpha_m$ *be non-zero algebraic numbers with degrees at most* d *and heights at most* A. *Further, let* $\beta_0, \ldots, \beta_m$ *be algebraic numbers with degrees at most* d *and heights at most* $B \geq 2$. *Then, either*

$$\Lambda := \beta_0 + \beta_1\log\alpha_1 + \cdots + \beta_m\log\alpha_m$$

equals zero or $|\Lambda| > B^{-C}$ *where* C *is an effectively computable constant depending only on* m, d, A *and the original determinations of the logarithms.*

The estimate for C takes the form $C'(\log A)^\kappa$ where κ depends only on m and C' depend only on m and d. Let us note that the special case of $m = 1$ of the above theorem leads to results of the form

$$|\log\alpha - \beta| > B^{-C}$$

for any algebraic number α not zero or 1 and for all algebraic numbers of degree at most d and heights at most $B \geq 2$. Here, C depends only on d and α. In particular, we can derive results of the form

$$|\pi - \beta| > B^{-C}.$$

Indeed, N.I. Feldman had already established the above inequality with C of the order of $d\log d$.

Further, when we restrict to the case when β is a rational number, these inequalities can be refined. For instance, for π we have the following lower bound

$$|\pi - p/q| > q^{-42}$$

for all rationals p/q ($q \geq 2$). This was established by Mahler. On the other hand, we have the following lower bound

$$|e^\pi - p/q| > q^{-c\log\log q}$$

for all rationals p/q and where c is an absolute constant. This was proved by Baker.

In 1993, Baker and Wüstholz [11] proved a sharper form of these theorems by offering a quantitative version of Baker's original theorem. We state a special case of their theorem.

Theorem 20.6 ([11]) *If $\beta_0 = 0$ and $\beta_1, \ldots, \beta_m$ are integers of absolute value at most B, then either $\Lambda = 0$ or* .

$$|\Lambda| > \exp(-(16md)^{2m+4}(\log^m A)\log B).$$

Finding sharp lower bounds for linear forms in logarithms of algebraic numbers constitutes an important theme in transcendence theory. We refer to the interested reader the book of Baker [9] and the recent monograph of Baker and Wüstholz [12] for further details.

We now apply Baker's theory to the study of $L(1, \chi)$ where $L(s, \chi)$ is the classical Dirichlet L-function attached to a non-trivial character χ. This is a prelude to the theme of applying Baker's theory to more general Dirichlet series which we take up in later chapters.

Let χ be a non-trivial Dirichlet character mod q with $q > 1$. For $s \in \mathbb{C}$ with $\text{Re}(s) > 1$, let

$$L(s, \chi) = \sum_{n=1}^{\infty} \frac{\chi(n)}{n^s}$$

be the associated Dirichlet L-function. It is classical that $L(s, \chi)$ extends to an entire function and that

$$L(1, \chi) = \sum_{n=1}^{\infty} \frac{\chi(n)}{n}.$$

Furthermore, $L(1, \chi) \neq 0$ by a theorem of Dirichlet. We are interested in the algebraic nature of $L(1, \chi)$. Now for any such χ, let

$$\hat{\chi}(n) = \frac{1}{q} \sum_{m=1}^{q} \chi(m) e^{-2\pi i m n/q}$$

be its Fourier transform. By orthogonality, we have

$$\chi(n) = \sum_{m=1}^{q} \hat{\chi}(m) e^{2\pi i m n/q}.$$

Note that $\hat{\chi}(q) = 0$. Now we are ready to prove the following:

Theorem 20.7 *If χ is a non-trivial Dirichlet character mod q with $q > 1$, then $L(1, \chi)$ is transcendental.*

Proof. By the previous discussions, we have

$$L(1, \chi) = \sum_{n=1}^{\infty} \frac{\chi(n)}{n}$$

$$= \sum_{n=1}^{\infty} \frac{1}{n} \sum_{m=1}^{q-1} \hat{\chi}(m) e^{2\pi i m n/q}$$

$$= -\sum_{m=1}^{q-1} \hat{\chi}(m) \log(1 - e^{2\pi im/q}).$$

This is a non-zero linear form in logarithms of algebraic numbers with algebraic coefficients. By Baker's theorem, this is transcendental. □

We end the chapter with one of the major applications of Baker's theory which is the explicit determination of all imaginary quadratic fields with class number one. This problem has a venerable history. We recommend the expository article of Goldfeld [52] and the recent monograph of Baker and Wüstholz [12] for a more detailed account of this topic.

Gauss conjectured that the only imaginary quadratic fields $\mathbb{Q}(\sqrt{-d})$, with $d > 0$ and squarefree, that have class number one are given by

$$d = 1, 2, 3, 7, 11, 19, 43, 67, 163.$$

In 1967, Baker [8] and Stark [117] independently solved this conjecture. We indicate below the main features of Baker's argument using linear forms in logarithms.

We shall be needing some familiarity with algebraic number theory. We suggest [76, 113] as possible references. Recall that if $k/\mathbb{Q}$ is a quadratic extension, its Dedekind zeta function $\zeta_k(s)$ factors as

$$\zeta_k(s) = \zeta(s)L(s, \chi) \qquad (*)$$

where χ is a quadratic Dirichlet character. In fact if D is the discriminant of k, then we may write $k = \mathbb{Q}(\sqrt{D})$ and $\chi(n) = \left(\frac{D}{n}\right)$ is the Kronecker symbol. The class number formula of Dirichlet can be stated as follows. If k is an imaginary quadratic field, then

$$L(1, \chi) = \frac{2\pi h(k)}{\omega_k \sqrt{|D|}}, \qquad D < 0$$

and if k is a real quadratic field,

$$L(1, \chi) = \frac{2h(k) \log \epsilon_k}{\sqrt{D}}, \qquad D > 0$$

where $h(k)$ denotes the class number of k and ϵ_k is the fundamental unit of k and ω_k is the number of roots of unity in k. We may write $(*)$ in another way, using zeta functions attached to binary quadratic forms. Given a form

$$f(x, y) = ax^2 + bxy + cy^2$$

with discriminant $D = b^2 - 4ac < 0$, we may associate the following function

$$\zeta(s, f) = \sum_{m,n}' \frac{1}{f(m,n)^s},$$

where the dash indicates $(m, n) \neq (0, 0)$. One can show that $\zeta(s, f)$ extends to the entire complex plane, apart from a simple pole at $s = 1$. Kronecker's limit formula explicitly gives the residue and the constant term in the Laurent expansion of $\zeta(s, f)$ at $s = 1$. Using the standard equivalence between binary quadratic forms of discriminant D and ideal classes of k, we may write $(*)$ as

$$\zeta_k(s) = \frac{1}{2} \sum_f \zeta(s, f)$$

where the sum is over a complete set of inequivalent quadratic forms with discriminant D. In other words,

$$\zeta(s)L(s, \chi) = \frac{1}{2} \sum_f \sum_{m,n}{}' \frac{1}{f(m, n)^s}.$$

We may twist this by a Dirichlet character χ_1 to get

$$L(s, \chi_1)L(s, \chi\chi_1) = \frac{1}{2} \sum_f \sum_{m,n}{}' \frac{\chi_1(f(m, n))}{f(m, n)^s}.$$

By classical theory, the inner sums are Mellin transforms of modular forms of weight one. The behaviour at $s = 1$ of the inner sum can be determined by Kronecker's second limit formula (see [77], for instance) when χ_1 is non-trivial. We are especially interested in applying this for the case

$$\chi(n) = \left(\frac{D}{n}\right) \qquad \text{and} \qquad \chi_1(n) = \left(\frac{D_1}{n}\right)$$

with $D_1 > 0$. Using the limit formula, we get

$$L(1, \chi_1)L(1, \chi\chi_1) = \frac{\pi^2}{6} \prod_{p|D_1} \left(1 - \frac{1}{p^2}\right) \sum_f \frac{\chi_1(a)}{a} + \sum_f \sum_{r=-\infty}^{\infty} A_r e^{\pi i r b/D_1 a} \qquad (**).$$

Here for $r \neq 0$,

$$|A_r| \leq \frac{2\pi|r|}{\sqrt{|D|}} e^{-s/aD_1}$$

with $s = \frac{\pi|r|}{\sqrt{|D|}}$. As regards A_0, it is equal to zero if D_1 is not a prime power. On the other hand if D_1 is a power of a prime p, then

$$A_0 = -\frac{2\pi\chi_1(a)}{D_1\sqrt{|D|}} \log p.$$

Now suppose $\mathbb{Q}(\sqrt{D})$ has class number one. Then by genus theory (see for example, [113]), if $-D > 2$, then $-D \equiv 3 \pmod{4}$ and is necessarily a prime. Moreover as the class number is one, there is only one form (up to equivalence) and which we can take to be

$$x^2 + xy + \left(\frac{1 - D}{4}\right) y^2.$$

By Dirichlet's class number formula,

$$L(1, \chi_1) = \frac{2h_1 \log \epsilon_1}{\sqrt{D_1}}$$

where h_1 is the class number of $\mathbb{Q}(\sqrt{D_1})$ and ϵ_1 is the fundamental unit attached to this real quadratic field. The quadratic character $\chi\chi_1$ corresponds to the imaginary quadratic field

$$\mathbb{Q}(\sqrt{DD_1})$$

and we have

$$L(1, \chi\chi_1) = \frac{h_2 \pi}{\sqrt{|DD_1|}}$$

where h_2 is the class number of $\mathbb{Q}(\sqrt{|DD_1|})$.

We will choose D_1 appropriately. Assuming $|D| > D_1$ so that $(D, D_1) = 1$, we obtain from (**),

$$\left| 2h_1 h_2 \log \epsilon_1 - \frac{\pi}{6} D_1 \sqrt{|D|} \prod_{p|D_1} \left(1 - \frac{1}{p^2} \right) \right| \leq \frac{D_1 \sqrt{|D|}}{\pi} \sum_{r=-\infty}^{\infty} |A_r|.$$

If we choose D_1 such that it is not a prime power, we are ensured $A_0 = 0$. We will choose $D_1 = 21$ and $D_1 = 33$ and in both cases $\mathbb{Q}(\sqrt{D_1})$ has class number one. Using the upper bounds for $|A_r|$, we obtain for $|D|$ large enough and $D_1 = 21$,

$$\left| h_2 \log \epsilon_2 - \frac{32}{21} \pi \sqrt{|D|} \right| < e^{-\pi\sqrt{|D|}/100}$$

where h_2 is the class number of $\mathbb{Q}(\sqrt{21D})$ and ϵ_2 is the fundamental unit of $\mathbb{Q}(\sqrt{21})$. Similarly for $D_1 = 33$, we obtain

$$\left| h_3 \log \epsilon_3 - \frac{80}{33} \pi \sqrt{|D|} \right| < e^{-\pi\sqrt{|D|}/100}$$

where h_3 is the class number of $\mathbb{Q}(\sqrt{33D})$ and ϵ_3 is the fundamental unit of $\mathbb{Q}(\sqrt{33})$. By eliminating the $\pi\sqrt{|D|}$ term, we obtain

$$|35h_2 \log \epsilon_2 - 22h_3 \log \epsilon_3| < 57e^{-\pi\sqrt{|D|}/100}.$$

The terms h_2 and h_3 can be bounded effectively by an inequality of the form

$$h_2, h_3 < c_1 \sqrt{|D|} \log |D|$$

with c_1 effectively computable. By Baker's theory on lower bounds for linear forms in logarithms, we have

$$|35h_2 \log \epsilon_2 - 22h_3 \log \epsilon_3| > B^{-c}$$

where $B = \max(35h_2, 22h_3)$ and c is an effectively computable constant (dependent on ϵ_2 and ϵ_3). We obtain

$$e^{-c_2 \log |D|} < |35h_2 \log \epsilon_2 - 22h_3 \log \epsilon_3| < 57 e^{-\pi \sqrt{|D|}/100}$$

from which $|D|$ is effectively bounded. One explicitly determines a bound for $|D|$ which is of the order of 10^{500}.

Heilbronn and Linfoot had previously shown that there are at most ten imaginary quadratic fields with class number one and Lehmer had given a lower bound for the tenth fictitious prime $p > 163$ such that $\mathbb{Q}(\sqrt{-p})$ has class number one. This lower bound was of the order of 10^9. Later, this lower bound was improved by Stark [116] who showed that the lower bound is of the order of e^{10^7}. Thus comparing with the upper bound obtained before, the classification of all imaginary quadratic fields with class number one is done.

This method extends to determine effectively all imaginary quadratic fields with class number two and has been carried out by Baker [8] and Stark [117]. There are precisely 18 such fields.

In 1976, Goldfeld [50, 51] used the theory of elliptic curves to obtain an effective lower bound for the class number of an imaginary quadratic field. But his proof was conditional upon the existence of an elliptic curve of Mordell–Weil rank 3 and whose associated L-series has a zero of order 3. In 1983, Gross and Zagier [56] found such an elliptic curve. Combining this with Goldfeld's result led to the following: for every $\epsilon > 0$, there is an effectively computable constant $c > 0$ such that the class number of $\mathbb{Q}(\sqrt{D})$ is greater than $c(\log |D|)^{1-\epsilon}$. In 1984, Oesterlé [90] refined the argument to give the lower bound

$$\frac{1}{7000}(\log |D|) \prod_{\substack{p | D \\ p \neq |D|}} \left(1 - \frac{[\, 2\sqrt{p} \,]}{p+1}\right)$$

for the class number of the imaginary quadratic field $\mathbb{Q}(\sqrt{D})$.

The scenario for class numbers of real quadratic fields is expected to be different. It is conjectured that there are infinitely many real quadratic fields with class number one. However, we do not even know if there are infinitely many number fields with class number one.

Exercises

1. Let $P(x)$ be a polynomial of degree $r \geq 2$. Assume that $P(x)$ has algebraic coefficients and that all of its roots are rational and not integral. Show that

$$\sum_{n=1}^{\infty} 1/P(n)$$

is either zero or transcendental. [Hint: Consider the partial fraction decomposition of $1/P(x)$.]

2. Show that the conclusion of the previous exercise is still valid for the sum

$$\sum_{n=1}^{\infty} Q(n)/P(n)$$

where $Q(x)$ is also a polynomial with algebraic coefficients and the degree of $Q(x)$ is at most $r - 2$.

3. Let f be an algebraic-valued function defined on the integers. Suppose that for some natural number $q > 1$, we have $f(n + q) = f(n)$ for all natural numbers n. Suppose further that

$$\sum_{a=1}^{q} f(a) = 0.$$

Show that

$$\sum_{n=1}^{\infty} \frac{f(n)}{n}$$

converges and that it is either zero or a transcendental number.

4. Suppose that the sum

$$F(z; x) := \sum_{n=1}^{\infty} \frac{z^n}{n + x}$$

converges. If z is algebraic and x is rational, show that the sum is either zero or a transcendental number.

Chapter 21

Schanuel's Conjecture

One of the most far reaching conjectures in transcendence theory is the following due to S. Schanuel:

Schanuel's Conjecture: Suppose $\alpha_1, \ldots, \alpha_n$ are complex numbers which are linearly independent over $\mathbb{Q}$. Then the transcendence degree of the field

$$\mathbb{Q}(\alpha_1, \ldots, \alpha_n, e^{\alpha_1}, \ldots, e^{\alpha_n})$$

over $\mathbb{Q}$ is at least n.

This conjecture is believed to include all known transcendence results as well as all reasonable transcendence conjectures on the values of the exponential function. Note that when the α_i's are algebraic numbers, this is the Lindemann–Weierstrass theorem.

In this chapter, we derive some interesting consequences of this conjecture. We begin with the following special case of Schanuel's conjecture. This generalises Baker's theorem. Let us refer to it as the weak Schanuel's conjecture.

Weak Schanuel's Conjecture: Let $\alpha_1, \ldots, \alpha_n$ be non-zero algebraic numbers such that $\log \alpha_1, \ldots, \log \alpha_n$ are linearly independent over $\mathbb{Q}$. Then these numbers are algebraically independent.

This special version itself has strong ramifications. For instance, it suffices to derive transcendence of special values of a number of L-functions arising from various analytic and arithmetic contexts.

The following is an important consequence of the weak Schanuel's conjecture.

Theorem 21.1 *Assume the weak Schanuel's conjecture. Let $\alpha_1, \ldots, \alpha_n$ be non-zero algebraic numbers. Then for any polynomial $f(x_1, \ldots, x_n)$ with algebraic coefficients such that $f(0, \ldots, 0) = 0$, $f(\log \alpha_1, \ldots, \log \alpha_n)$ is either zero or transcendental.*

M.R. Murty and P. Rath, *Transcendental Numbers*, DOI 10.1007/978-1-4939-0832-5_21, 111
© Springer Science+Business Media New York 2014

Proof. We use induction on n. For $n = 1$, it is true by the classical Lindemann–Weierstrass theorem. Now for $n \geq 2$, let $f(x_1, \ldots, x_n)$ be a polynomial in $\overline{\mathbb{Q}}[x_1, \ldots, x_n]$ with $f(0, \ldots, 0) = 0$. Further, suppose that $A := f(\log \alpha_1, \ldots, \log \alpha_n)$ is algebraic. By the weak Schanuel's conjecture, the numbers $\log \alpha_1, \ldots, \log \alpha_n$ are linearly dependent over $\mathbb{Q}$. Then there exists integers $c_1, \ldots, c_n$ such that

$$c_1 \log \alpha_1 + \cdots + c_n \log \alpha_n = 0.$$

Suppose $c_1 \neq 0$. Then $\log \alpha_1 = -\frac{1}{c_1}(c_2 \log \alpha_2 + \cdots + c_n \log \alpha_n)$. Replacing this value of $\log \alpha_1$ in the expression for A, we have

$$A = g(\log \alpha_2, \ldots, \log \alpha_n),$$

where $g(x_1, \ldots, x_{n-1})$ is a polynomial with algebraic coefficients in $n - 1$ variables. Then by induction hypothesis $A = 0$. This completes the proof. $\square$

Now we proceed to derive some other consequences of Schanuel's conjecture:

Theorem 21.2 *Assume that Schanuel's conjecture is true. Let $\alpha \neq 0, 1$ be algebraic. Then $\log \alpha$ and $\log \log \alpha$ are algebraically independent.*

Proof. Note that for $\alpha \in \overline{\mathbb{Q}} \setminus \{0, 1\}$, $\log \alpha$ and $\log \log \alpha$ are linearly independent over $\mathbb{Q}$. We apply Schanuel's conjecture to the numbers $\log \alpha$ and $\log \log \alpha$. Then we see that the transcendence degree of the field $\mathbb{Q}(\log \alpha, \log \log \alpha, \alpha)$ is two and hence $\log \alpha$ and $\log \log \alpha$ are algebraically independent. $\square$

Theorem 21.3 *Assume that Schanuel's conjecture is true. If $\alpha_1, \ldots, \alpha_n \in \overline{\mathbb{Q}}$ are linearly independent over $\mathbb{Q}$, then $\pi, e^{\alpha_1}, \ldots, e^{\alpha_n}$ are algebraically independent. In particular, e and π are algebraically independent.*

Proof. We apply Schanuel's conjecture to the $\mathbb{Q}$-linearly independent numbers $\alpha_1, \ldots, \alpha_n$ and $i\pi$ to get the result. $\square$

Theorem 21.4 *Assume that Schanuel's conjecture is true. If $\alpha_1, \ldots, \alpha_n$ are algebraic numbers such that $i, \alpha_1, \ldots, \alpha_n$ are linearly independent over $\mathbb{Q}$, then $\pi, e^{\alpha_1 \pi}, \ldots, e^{\alpha_n \pi}$ are algebraically independent.*

Proof. Apply Schanuel's conjecture to the $\mathbb{Q}$-linearly independent numbers $i\pi, \quad \alpha_1 \pi, \quad \ldots, \quad \alpha_n \pi$ to get the result. $\square$

Thus Schanuel's conjecture implies that π and e^{π} are algebraically independent. This has been established unconditionally by Nesterenko.

Theorem 21.5 *Assume that Schanuel's conjecture is true. Then π^e is transcendental.*

Proof. By Nesterenko's result, we know that π and $\log \pi$ are linearly independent over $\mathbb{Q}$. We apply Schanuel's conjecture to the $\mathbb{Q}$-linearly independent numbers $1, i\pi$ and $\log \pi$ to conclude that e, π and $\log \pi$ are algebraically independent. Now apply Schanuel's conjecture to the $\mathbb{Q}$-linearly independent numbers $1, \log \pi, i\pi + e \log \pi, e \log \pi$. $\square$

Let us define a *Baker period* to be an element of the $\overline{\mathbb{Q}}$-vector space spanned by logarithms of non-zero algebraic numbers.

Theorem 21.6 *Assume that Schanuel's conjecture is true. If $\alpha_1, \ldots, \alpha_n$ are non-zero algebraic numbers such that $\log \alpha_1, \ldots, \log \alpha_n$ are linearly independent over $\mathbb{Q}$, then $\log \alpha_1, \ldots, \log \alpha_n, \log \pi$ are algebraically independent. In particular, $\log \pi$ is not a Baker period.*

Proof. Since $\log \alpha_1, \ldots, \log \alpha_n$ are linearly independent over $\mathbb{Q}$, the numbers $\log \alpha_1, \ldots, \log \alpha_n$ are algebraically independent by Schanuel's conjecture.

First suppose that $\pi, \log \alpha_1, \ldots, \log \alpha_n$ are linearly dependent over $\overline{\mathbb{Q}}$, i.e.

$$\pi = \beta_1 \log \alpha_1 + \cdots + \beta_n \log \alpha_n,$$

where $\beta_i \in \overline{\mathbb{Q}}$ and not all of them are zero. Without loss of generality, assume that $\beta_1 \neq 0$. Then $\pi, \log \alpha_2, \ldots, \log \alpha_n$ are linearly independent over $\overline{\mathbb{Q}}$. Now applying Schanuel's conjecture to the $\mathbb{Q}$-linearly independent numbers $i\pi, \log \alpha_2, \ldots, \log \alpha_n, \log \pi$ we see that $\log \alpha_1, \ldots, \log \alpha_n, \log \pi$ are algebraically independent.

Next suppose that π and $\log \alpha_1, \ldots, \log \alpha_n$ are linearly independent over $\overline{\mathbb{Q}}$. Then we apply Schanuel's conjecture to the $\mathbb{Q}$-linearly independent numbers $i\pi, \log \alpha_1, \ldots, \log \alpha_n, \log \pi$ to get the required result. $\square$

Theorem 21.7 *Assume that Schanuel's conjecture is true. If α is a non-zero Baker period, then $1/\alpha$ is not a Baker period. In particular, $1/\pi$ is not a Baker period.*

Proof. Since α is a Baker period, we can write

$$\alpha = \beta_1 \log \delta_1 + \cdots + \beta_n \log \delta_n,$$

where $\beta_i, \delta_i \in \overline{\mathbb{Q}} \setminus \{0\}$. If $1/\alpha$ is also a Baker period, then

$$\frac{1}{\alpha} = \gamma_1 \log \alpha_1 + \cdots + \gamma_k \log \alpha_k,$$

where $\gamma_i, \alpha_i \in \overline{\mathbb{Q}} \setminus \{0\}$. This implies that

$$1 = f\left(\log \delta_1, \ldots, \log \delta_n, \log \alpha_1, \ldots, \log \alpha_k\right), \tag{21.1}$$

where f is a polynomial in $\overline{\mathbb{Q}}[x_1, \ldots, x_{n+k}]$ with $f(0, \ldots, 0) = 0$. Then the right-hand side of (21.1) is either zero or transcendental and hence the result follows. $\square$

One can show that Schanuel's conjecture implies that e^e is transcendental (see Exercise 1). We refer to the papers of Waldschmidt [128] and Brownawell [22] for an interesting theorem in this context, namely that either e^e or e^{e^2} is transcendental. This was a conjecture of Schneider.

We note that Kontsevich and Zagier [74] have introduced the notion of periods. A *period* is a complex number whose real and imaginary parts are values of absolutely convergent integrals of rational functions with rational coefficients over domains in $\mathbb{R}^n$ given by polynomial inequalities with rational coefficients. Clearly all algebraic numbers are periods. On the other hand, π is a period for it is expressible as

$$\pi = \int \int_{x^2+y^2 \leq 1} dx dy.$$

Further, non-zero Baker periods are examples of transcendental periods. This follows from Baker's theorem.

The set of periods forms a ring. It is an open question to determine whether the group of units of this ring contains only the obvious units, namely the non-zero algebraic numbers. We shall come back to these periods in the last chapter.

We now apply Schanuel's conjecture to study some special values of the Gamma function.

Theorem 21.8 *For any rational number* $x \in (0, 1/2]$, *the number*

$$\log \Gamma(x) + \log \Gamma(1 - x)$$

is transcendental with at most one possible exception.

Proof. Using the reflection property of the gamma function, we have

$$\log \Gamma(x) + \log \Gamma(1 - x) = \log \pi - \log \sin \pi x.$$

If x_1 and x_2 are distinct rational numbers with

$$\log \Gamma(x_i) + \log \Gamma(1 - x_i) \in \overline{\mathbb{Q}}, \quad i = 1, 2,$$

then their difference $\log \sin \pi x_2 - \log \sin \pi x_1$ is an algebraic number. But this is a non-zero Baker period and hence transcendental. $\square$

The possible fugitive exception in the above theorem can be removed if we assume Schanuel's conjecture.

Theorem 21.9 *Schanuel's conjecture implies that*

$$\log \Gamma(x) + \log \Gamma(1 - x)$$

is transcendental for every rational $0 < x < 1$.

Proof. As noticed earlier, Schanuel's conjecture implies that for any non-zero algebraic number α, the two numbers e^α and π are algebraically independent. Suppose $\alpha = \log \Gamma(x) + \log \Gamma(1-x)$ is algebraic. Then $e^\alpha \sin(\pi x) = \pi$ which contradicts the algebraic independence of e^α and π. $\square$

We also have,

Theorem 21.10 *Schanuel's conjecture implies that for any rational $x \in (0,1)$, at least one of the following statement is true:*

1. *Both $\Gamma(x)$ and $\Gamma(1-x)$ are transcendental.*

2. *Both $\log \Gamma(x)$ and $\log \Gamma(1-x)$ are transcendental.*

Proof. If (1) is true, there is nothing to prove. Without loss of generality, suppose that $\Gamma(x)$ is algebraic for some $x \in \mathbb{Q}$. Then $\log \Gamma(x)$ is a Baker period. Since

$$\log \Gamma(1-x) = -\log \Gamma(x) + \log \pi - \log \sin x\pi,$$

therefore it follows that $\log \Gamma(1-x)$ is transcendental. $\square$

The logarithms of the gamma function as well as $\log \pi$ are of central importance in studying the special values of a general class of L-functions.

Finally, we now apply Schanuel's conjecture in the investigation of some special values of Dedekind zeta functions. We refer to [58] for a more detailed account. The relevant details from algebraic number theory can be found in the books of Lang [76] or Neukirch [88]. Let K be a number field of degree n. For $\Re(s) > 1$, the *Dedekind zeta function* of K is defined as

$$\zeta_K(s) = \sum_{\mathfrak{a}} \frac{1}{N(\mathfrak{a})^s},$$

where the sum is over all the integral ideals of $\mathcal{O}_K$, the ring of integers of K. When $K = \mathbb{Q}$, this is the Riemann zeta function. Analogous to the Riemann zeta function, $\zeta_K(s)$ is analytic for $\Re(s) > 1$ and $(s-1)\zeta_K(s)$ extends to an entire function with

$$\lim_{s \to 1^+} (s-1)\zeta_K(s) = \operatorname{Res}_{s=1}\zeta_K(s) = \frac{2^{r_1}(2\pi)^{r_2} h_K R_K}{w_K \sqrt{|d_K|}},$$

where r_1 is the number of real embeddings, $2r_2$ is the number of complex embeddings, h_K is the class number, R_K is the regulator (which is known to be non-zero), w_K is the number of roots of unity in K and d_K is the discriminant of K.

We are interested in the nature of $\operatorname{Res}_{s=1}\zeta_K(s)$ and the regulator R_K. Because of the presence of π, the transcendence of one does not imply the transcendence of the other unless K is a totally real field.

Theorem 21.11 *Assume the weak Schanuel's conjecture. Let K be a number field with unit rank at least 1. Then both the regulator R_K and $\mathrm{Res}_{s=1}\zeta_K(s)$ are transcendental.*

Proof. By the class number formula,

$$\mathrm{Res}_{s=1}\zeta_K(s) = \frac{2^{r_1}(2\pi)^{r_2} h_K R_K}{\omega \sqrt{|d_K|}},$$

where r_1 and $2r_2$ be the number of real and complex embeddings. Let $u^{(j)}$ be the j-th conjugate of $u \in K$ where j runs through the embeddings modulo complex conjugation. Let

$$\{u_1, u_2, \ldots, u_r\}$$

be a set of generators of the ordinary unit group modulo the roots of unity. Then the regulator R_K, up to an algebraic multiple, is given by

$$\begin{vmatrix} 1 & \log|u_1^{(1)}| & \cdots & \log|u_r^{(1)}| \\ \vdots & \vdots & \vdots & \vdots \\ 1 & \log|u_1^{(r+1)}| & \cdots & \log|u_r^{(r+1)}| \end{vmatrix}.$$

Clearly, by Theorem 21.1, the regulator R_K is transcendental.
Further,

$$\pi^{r_2} R_k = F\left(\log(-1), \log|u_1^{(1)}|, \ldots, \log|u_r^{(r+1)}|\right)$$

where F is a polynomial with algebraic coefficients whose constant term is zero. Assume that the weak Schanuel's conjecture is true. Then by Theorem 21.1, $\mathrm{Res}_{s=1}\zeta_K(s)$ is necessarily transcendental. $\square$

In the p-adic set-up, it is conjectured that the p-adic regulator rank of any number field K is equal to the rank of its unit group (see [89], for instance). This is referred to as *Leopoldt's conjecture.* Waldschmidt [129] has shown that the p-adic regulator rank is at least half of the expected value.

When K is a totally real field, then Leopoldt's conjecture is equivalent to the non-vanishing of the p-adic regulator of K (which is well-defined up to sign).

Leopoldt's conjecture is known to be true for abelian extensions $K|k$ where k is either $\mathbb{Q}$ or imaginary quadratic (see [5, 24]). The proof in the abelian case uses the p-adic analog of Baker's theorem and the notion of Dedekind determinants. We shall come across these determinants in a later chapter. The conjecture is open for arbitrary number fields.

We note that Leopoldt's conjecture is also related to the non-vanishing of special values of certain p-adic L-functions. We refer to the book of Washington [130] (see also the work of Colmez [35]) for further details.

One can extend these study to special values of Artin L-functions. The guideline for such an investigation is a program envisaged by Stark [118]. We refer to [95] for a more detailed account of Artin L-functions. Let K/k be Galois extension of number fields with Galois group $G = \mathrm{Gal}(K/k)$. Corresponding to any finite dimensional representation (ϕ, V) of G with character χ, the *Artin L-function* is defined by

$$L(s, \chi, K/k) = \prod_{\mathcal{P}} \det(1 - N(\mathcal{P})^{-s}\phi(\sigma_\beta)|_{V^{I_\beta}})^{-1}$$

where $\mathcal{P}$ runs over all the prime ideals in $\mathcal{O}_k$, β is a prime ideal lying over $\mathcal{P}$, I_β is its inertia group and σ_β is the associated Frobenius element in the Galois group. Stark [118] has made the following conjecture:

Conjecture (Stark). Suppose χ does not contain the trivial character χ_0 as a constituent. Then

$$L(1, \chi, K/k) = \frac{W(\bar{\chi})2^a\pi^b}{(|d_k|N(f))^{1/2}}\theta(\bar{\chi})R(\bar{\chi}).$$

We refer to the article of Stark for descriptions of the terms involved. Stark proved the above conjecture for all rational characters.

Theorem 21.12 *Assume that the weak Schanuel's conjecture is true. Then for any rational nontrivial irreducible character χ, $L(1, \chi, K/k)$ is transcendental.*

Proof. Let χ be a character as above. Then as proved by Stark

$$L(1, \chi, K/k) = \frac{W(\bar{\chi})2^a\pi^b}{(|d_k|N(f))^{1/2}}\theta(\bar{\chi})R(\bar{\chi}).$$

In the expression on the right-hand side, there are two possible transcendental objects, namely π^b and $R(\chi)$. But we have a description of the number $R(\chi)$. It is the determinant of an a by a matrix whose entries are linear forms in logarithms of absolute values of units in K and its conjugate fields. For instance, when k is equal to $\mathbb{Q}$, the entries of this matrix are given by $c_{ij} = \sum_{\sigma \in G} a_{ij}(\sigma) \log(|\epsilon^\sigma|)$ where $A(\sigma) = (a_{ij}(\sigma))$ is a representation of G whose character is χ and ϵ is a Minkowski unit. Since $\log(-1) = i\pi$, the residue is the value of a polynomial of the form mentioned in Theorem 21.1 evaluated at logarithms of algebraic numbers. It is classical that for any irreducible character χ of G, for all $t \in \mathbb{R}$, $L(1 + it, \chi, K/k) \neq 0$ and hence by appealing to Theorem 21.1, we see that $L(1, \chi, K/k)$ is transcendental under the weak Schanuel's conjecture. $\square$

For more details and other applications to transcendence of Petersson norms of certain weight one modular forms, the reader may consult [58].

We end this chapter by mentioning a generalisation of Schanuel's conjecture. Let us first set up the preamble which motivates such a generalisation.

The conjecture of Schanuel is about the algebraic independence of the values of the exponential function. Analogous to the exponential set-up, there has also been progress in the elliptic world (see the survey article [127], for instance).

For a Weierstrass $\wp$-function with algebraic invariants g_2, g_3 and field of endomorphisms k, the following set

$$\mathcal{L}_E = \{\alpha \in \mathbb{C} \ : \ \wp(\alpha) \in \overline{\mathbb{Q}} \cup \{\infty\}\}$$

is referred to as the set of elliptic logarithms of algebraic points on E. Here E is the associated elliptic curve. Let Ω be the lattice of periods. This k-linear space $\mathcal{L}_E$ is the elliptic analog of the $\mathbb{Q}$-linear space of logarithms of non-zero algebraic numbers for the exponential case. The question of linear independence of elliptic logarithms, analogous to Baker's theorem, has been established by Masser for the CM case [83] and Bertrand and Masser for the non-CM case [18].

The algebraic independence of the values of the Weierstrass $\wp$-function is more delicate. When the Weierstrass $\wp$-function has complex multiplication, the following analogue of the Lindemann–Weierstrass theorem has been proved by Philippon [91] and Wüstholz [133].

Theorem 21.13 *(Philippon/Wüstholz) Let $\wp$ be a Weierstrass $\wp$-function with algebraic invariants g_2 and g_3 that has complex multiplication. Let k be its field of endomorphisms. Let*

$$\alpha_1, \alpha_2, \ldots, \alpha_n$$

be algebraic numbers which are linearly independent over k. Then the numbers $\wp(\alpha_1), \ldots, \wp(\alpha_n)$ are algebraically independent.

For the non-CM case, so far only the algebraic independence of at least $n/2$ of these numbers is known by the work of Chudnovsky [33].

In his seminal work, Nesterenko proved the following general result (see [86, Chap. 3, Corollary 1.6]) which involves both exponential and elliptic functions.

Proposition 21.14 *Let $\wp$ be a Weierstrass $\wp$-function with algebraic invariants g_2 and g_3 and with complex multiplication by an order in the field k. If ω is any period of $\wp$, η the corresponding quasi-period and τ is any element of k which is not real, then each of the sets*

$$\{\pi, \omega, e^{2\pi i \tau}\} \qquad \{\pi, \eta, e^{2\pi i \tau}\}$$

is algebraically independent.

With these background in mind, the following elliptic-exponential extension of the conjecture of Schanuel has been suggested in [62].

Conjecture: Let $\wp$ be a Weierstrass $\wp$-function with algebraic invariants g_2 and g_3 and lattice Ω. Let k be its field of endomorphisms. Let

$$\alpha_1, \alpha_2, \ldots, \alpha_r, \alpha_{r+1}, \ldots \alpha_n$$

*be complex numbers which are linearly independent over k and are not in Ω.
Then the transcendence degree of the field*

$$\mathbb{Q}(\alpha_1, \alpha_2, \ldots, \alpha_n, e^{\alpha_1}, \ldots, e^{\alpha_r}, \wp(\alpha_{r+1}) \ldots, \wp(\alpha_n))$$

over $\mathbb{Q}$ is at least n.

This conjecture is a special case of a more general conjecture formulated by Bertolin [16] (one needs to specialise "conjecture elliptico-torique" on p. 206 of [16] to the case of a single elliptic curve). We will also come across another elliptic-exponential extension of Schanuel's conjecture in Chap. 26.

It is worthwhile to mention that Schanuel also formulated an analogous conjecture for formal power series. This conjecture was proved by Ax [6] in 1971 which is the following:

Theorem 21.15 *(J. Ax) Let $y_1, \ldots, y_n \in t\mathbb{C}[[t]]$ be n formal power series which are linearly independent over $\mathbb{Q}$. Then the field extension*

$$\mathbb{C}(t)(y_1, \ldots, y_n, \exp(y_1), \ldots, \exp(y_n))$$

has transcendence degree at least n over $\mathbb{C}(t)$.

Furthermore, in the same paper, Ax considers the following conjecture:

Conjecture: Let $y_1, \ldots, y_n \in \mathbb{C}[[t_1, \ldots, t_m]]$ be $\mathbb{Q}$-linearly independent. Then the transcendence degree of the field

$$\mathbb{Q}(y_1, \ldots, y_n, \exp(y_1), \ldots, \exp(y_n))$$

over $\mathbb{Q}$ is at least $n + r$, where r is the rank of the $n \times m$ matrix $\left(\frac{\partial y_i}{\partial t_j}\right)$.

Clearly, the original Schanuel's conjecture involving complex numbers is a special case of the above. But Ax showed that the Schanuel's conjecture is actually equivalent to the above conjecture (see also [34]). Elliptic versions of Ax's results have been obtained by Brownawell and Kubota [23].

We end by noting that D. Roy has suggested an alternate algebraic approach towards the weak Schanuel's conjecture which is about the algebraic independence of logarithms of algebraic numbers.

Let us first formulate the following homogeneous version of the weak Schanuel's conjecture: Let $\alpha_1, \ldots, \alpha_n$ be non-zero algebraic numbers such that the numbers $\log \alpha_1, \ldots, \log \alpha_n$ are linearly independent over $\mathbb{Q}$. Then for any non-zero homogeneous polynomial $P(X_1, \ldots, X_n)$ with rational coefficients, $P(\log \alpha_1, \ldots, \log \alpha_n)$ is not equal to zero.

Now let M be an $m \times n$ matrix (λ_{ij}) where each λ_{ij} is the logarithm of a non-zero algebraic number for $1 \leq i \leq m, 1 \leq j \leq n$. For each such matrix, let V be the $\mathbb{Q}$-vector space generated by the mn entries of the above matrix. Let r be the dimension of this space and let $\{e_1, \ldots, e_r\}$ be a basis. Then

$$\lambda_{ij} = \sum_{k=1}^{r} b_{ijk} e_k,$$

where $b_{ijk} \in \mathbb{Q}$ and hence the matrix M is given by

$$\begin{vmatrix} \sum_{k=1}^{r} b_{11k}e_k & \cdots & \sum_{k=1}^{r} b_{1nk}e_k \\ \vdots & \vdots & \vdots \\ \sum_{k=1}^{r} b_{m1k}e_k & \cdots & \sum_{k=1}^{r} b_{mnk}e_k \end{vmatrix}.$$

We now consider the following formal matrix $M_{(for)}$ given by

$$\begin{vmatrix} \sum_{k=1}^{r} b_{11k}X_k & \cdots & \sum_{k=1}^{r} b_{1nk}X_k \\ \vdots & \vdots & \vdots \\ \sum_{k=1}^{r} b_{m1k}X_k & \cdots & \sum_{k=1}^{r} b_{mnk}X_k \end{vmatrix}$$

with entries in the field $\mathbb{Q}(X_1, \ldots, X_r)$ where $X_1, \ldots, X_r$ are variables.

The rank of this formal matrix $M_{(for)}$ associated with the matrix M is referred to as the *structural rank* of M and is independent of the choice of basis of V.

It is clear that the *homogeneous weak Schanuel's conjecture* implies that the rank of the matrix M is equal to its structural rank. D. Roy proved that the converse also holds. Here is Roy's theorem.

Theorem 21.16 *(D. Roy) Suppose that for any matrix M with entries in logarithms of nonzero algebraic numbers, rank of M is equal to its structural rank. Then the homogeneous weak Schanuel's conjecture is true.*

In this connection, Roy also proved that for any such matrix M, rank of M is at least half of its structural rank. Furthermore, by working with matrices having entries in the $\mathbb{Q}$-vector space generated by 1 and logarithms of algebraic numbers, one can link the structural rank of such matrices to the weak Schanuel's conjecture. See Chap. 12 of [125] for a detailed discussion about these results. We note that Roy (see [107]) has also suggested an alternate algebraic approach towards the original Schanuel's conjecture.

Finally, Schanuel's conjecture has been found to have implications in other contexts like model theory and commensurability of locally symmetric spaces (see [93, 136] for instance).

Exercises

1. Show that Schanuel's conjecture implies the four exponentials conjecture.

2. Assuming Schanuel's conjecture, show that e^e is transcendental.

3. Assuming Schanuel's conjecture, and **not** using Nesterenko's theorem, show that π and $\log \pi$ are algebraically independent.

4. Let $\beta_1, \ldots, \beta_n$ be linearly independent algebraic numbers over the rationals. Suppose that $\alpha_1, \ldots, \alpha_m$ are algebraic numbers such that

$\log \alpha_1, \ldots, \log \alpha_m$ are linearly independent over $\mathbb{Q}$. Assuming Schanuel's conjecture, show that

$$e^{\beta_1}, \ldots, e^{\beta_n}, \log \alpha_1, \ldots, \log \alpha_m$$

are algebraically independent over the rationals.

5. Let $z_1, \ldots, z_n$ be complex numbers. Show that $z_1, \ldots, z_n$ are algebraically independent over the rationals if and only if they are algebraically independent over the field of algebraic numbers.

6. Define a sequence of numbers E_n recursively as follows: $E_0 = 1$ and $E_n = \exp(E_{n-1})$ for $n \geq 1$. Show for any finite subset A of the natural numbers, the set of numbers E_a with $a \in A$ is an algebraically independent set of numbers, assuming Schanuel's conjecture.

7. Define a sequence of number P_n recursively as follows: $P_0 = \pi$, and $P_n = \pi^{P_{n-1}}$ for $n \geq 1$. Assuming Schanuel's conjecture, show that any finite subset of the set of P_n's is algebraically independent over the rationals.

8. Show that if $\alpha_1, \ldots, \alpha_n$ are $\mathbb{Q}$-linearly independent algebraic numbers, then Schanuel's conjecture is true. (This is a consequence of the Lindemann–Weierstrass theorem. See Theorem 4.1 in Chap. 4.)

Chapter 22

Transcendental Values of Some Dirichlet Series

There is a large collection of Dirichlet series defined purely arithmetically that have been conjectured to have analytic continuation and functional equations. Deligne [38] has formulated a far-reaching conjecture regarding the special values of these series at special points in the complex plane and one would like to know if these special values are transcendental numbers or not. The most notable example is the L-function attached to an elliptic curve and the Birch and Swinnerton-Dyer conjecture. In a lecture at the Stony Brook conference on number theory in the summer of 1969, Sarvadaman Chowla posed the following question. Does there exist a rational-valued arithmetic function f, periodic with prime period p such that

$$\sum_{n=1}^{\infty} \frac{f(n)}{n}$$

converges and equals zero? In 1973, Baker, Birch and Wirsing ([10], see also [29], [31] and [101]) answered this question in the following theorem:

Theorem 22.1 *If f is a non-zero function defined on the integers with algebraic values and period q such that $f(n) = 0$ whenever $1 < (n, q) < q$ and the q-th cyclotomic polynomial is irreducible over $\mathbb{Q}(f(1), \ldots, f(q))$, then*

$$\sum_{n=1}^{\infty} \frac{f(n)}{n} \neq 0.$$

In particular, if f is rational valued, the second condition holds trivially. If q is prime, then the first condition is vacuous. Thus, the theorem resolves Chowla's question. We shall present a proof of this theorem in the next chapter. In 2001, Adhikari et al. [2] noted that the theory of linear forms in logarithms

M.R. Murty and P. Rath, *Transcendental Numbers*, DOI 10.1007/978-1-4939-0832-5_22, 123
© Springer Science+Business Media New York 2014

can be used to show that in fact, the sum in the above theorem is transcendental whenever it converges.

Let us first derive a necessary and sufficient condition for the sum in the above theorem to converge. To this end, we use the Hurwitz zeta function. Recall that for real x with $0 < x \leq 1$, this function is defined as the series

$$\zeta(s,x) := \sum_{n=0}^{\infty} \frac{1}{(n+x)^s},$$

for $\Re(s) > 1$. Note that the series $\zeta(s,1)$ is the familiar Riemann zeta function. Hurwitz (see [4], for instance) proved that this function extends meromorphically to the complex plane with a simple pole at $s = 1$ and residue 1. Moreover, we have the following important fact:

$$\lim_{s \to 1^+} \zeta(s,x) - \frac{1}{s-1} = -\frac{\Gamma'(x)}{\Gamma(x)}.$$

This is easily seen as follows:

$$\lim_{s \to 1^+} \zeta(s,x) - \zeta(s) = \frac{1}{x} + \sum_{n=1}^{\infty} \left(\frac{1}{n+x} - \frac{1}{n} \right).$$

From the Hadamard factorisation of $1/\Gamma(z)$,

$$\frac{1}{\Gamma(z)} = z e^{\gamma z} \prod_{n=1}^{\infty} \left(1 + \frac{z}{n} \right) e^{-z/n},$$

we have by logarithmic differentiation,

$$-\frac{\Gamma'}{\Gamma}(z) = \gamma + \frac{1}{z} + \sum_{n=1}^{\infty} \left(\frac{1}{n+z} - \frac{1}{n} \right).$$

Thus,

$$\lim_{s \to 1^+} \zeta(s,x) - \zeta(s) = -\gamma - \frac{\Gamma'}{\Gamma}(x).$$

Observe that in the special case that $x = 1$, we deduce that

$$\Gamma'(1) = -\gamma.$$

Recall that by partial summation, we have

$$\zeta(s) = s \int_1^{\infty} \frac{[x]}{x^{s+1}} dx.$$

The integral can be written as

$$\frac{s}{s-1} - s \int_1^{\infty} \frac{\{x\}}{x^{s+1}} dx,$$

so that

$$\lim_{s \to 1+} \zeta(s) - \frac{1}{s-1} = 1 - \int_1^\infty \frac{\{x\}}{x^2} dx.$$

This last integral is easily evaluated as

$$\lim_{N \to \infty} \int_1^N \frac{\{x\}}{x^2} dx = \lim_{N \to \infty} \left(\log N - \sum_{n=1}^{N-1} n \int_n^{n+1} \frac{dx}{x^2} \right)$$

$$= \lim_{N \to \infty} \left(\log N - \sum_{n=2}^N \frac{1}{n} \right) = 1 - \gamma.$$

Thus,

$$\lim_{s \to 1+} \zeta(s) - \frac{1}{s-1} = \gamma.$$

Putting everything together, we obtain

Theorem 22.2

$$\lim_{s \to 1+} \zeta(s, x) - \frac{1}{s-1} = -\frac{\Gamma'(x)}{\Gamma(x)}.$$

Let f be any periodic arithmetic function with period q, that is

$$f : \mathbb{N} \to \mathbb{C} \quad \text{such that} \quad f(n+q) = f(n) \quad \forall n.$$

Then for $s \in \mathbb{C}$ with $\Re(s) > 1$, let $L(s, f)$ be defined as

$$L(s, f) = \sum_{n=1}^\infty \frac{f(n)}{n^s}.$$

Now running over arithmetic progressions modulo q, we have the following expression:

$$L(s, f) = q^{-s} \sum_{a=1}^q f(a) \zeta(s, a/q), \quad \Re(s) > 1.$$

We can write this expression as

$$L(s, f) = q^{-s} \sum_{a=1}^q f(a) \left[\zeta(s, a/q) - \frac{1}{s-1} \right] + \frac{q^{-s}}{s-1} \sum_{a=1}^q f(a).$$

This and partial summation yields the following theorem.

Theorem 22.3 *Let f be any periodic arithmetic function with period q. Then the series*

$$\sum_{n=1}^\infty \frac{f(n)}{n}$$

converges if and only if

$$\sum_{a=1}^{q} f(a) = 0,$$

and in the case of convergence, the value of the series is

$$-\frac{1}{q}\sum_{a=1}^{q} f(a)\frac{\Gamma'}{\Gamma}(a/q).$$

This gives us an interesting corollary even in the classical case:

Theorem 22.4 *For a non-trivial character χ mod q,*

$$L(1,\chi) = -\frac{1}{q}\sum_{a \bmod q} \chi(a)\frac{\Gamma'}{\Gamma}(a/q).$$

Let us now come back to the type of functions considered in the theorem of Baker, Birch and Wirsing. We are now ready to analyse the series:

$$\sum_{n=1}^{\infty}\frac{f(n)}{n^s} = q^{-s}\sum_{a=1}^{q} f(a)\zeta(s,a/q).$$

By the previous theorem, we have that

$$\sum_{a=1}^{q} f(a) = 0$$

is a necessary and sufficient condition for the convergence of the series at $s = 1$. So let us assume that f satisfies the above. Then we have the following expression:

$$\sum_{n=1}^{\infty}\frac{f(n)}{n} = -\frac{1}{q}\sum_{a=1}^{q} f(a)\frac{\Gamma'}{\Gamma}(a/q).$$

We now try to derive an alternate expression for the above series as we did for $L(1,\chi)$ earlier. As before for any such periodic function f with period q, let

$$\hat{f}(n) = \frac{1}{q}\sum_{m=1}^{q} f(m)e^{-2\pi imn/q}$$

be its Fourier transform. By orthogonality, we have

$$f(n) = \sum_{m=1}^{q} \hat{f}(m)e^{2\pi imn/q}.$$

Note that

$$\hat{f}(q) = 0 \quad \text{as} \quad \sum_{a=1}^{q} f(a) = 0.$$

Carrying out the explicit evaluation for $L(1, f)$ as done earlier for $L(1, \chi)$, we immediately have the following:

Theorem 22.5 *Let f be any function defined on the integers and with period q. Assume further that*

$$\sum_{a=1}^{q} f(a) = 0.$$

Then,

$$\sum_{n=1}^{\infty} \frac{f(n)}{n} = -\frac{1}{q} \sum_{a=1}^{q} f(a) \frac{\Gamma'}{\Gamma}(a/q)$$

$$= -\sum_{m=1}^{q-1} \hat{f}(m) \log(1 - e^{2\pi i m/q}).$$

Thus, in particular, if f takes algebraic values, the series is either zero or transcendental.

Thus we see that when f takes algebraic values, then the above series is a linear form in logarithms of algebraic numbers for which Baker's theorem applies. In particular, it is either zero or transcendental. The former case is ruled out in the case when f is rational-valued and $f(a)$ is equal to zero for $1 < (a, q) < q$. This is by the theorem of Baker, Birch and Wirsing which we shall derive in the next chapter. However, this observation allows us to deduce the following:

Theorem 22.6 *Let $q > 1$. At most one of the $\phi(q)$ values*

$$\frac{\Gamma'}{\Gamma}(a/q), \quad 1 \le a < q, \quad (a, q) = 1,$$

is algebraic.

Proof. If we choose two distinct residue classes $a, b \bmod q$, and set $f(a) = 1$, $f(b) = -1$ with f zero otherwise, then, f satisfies the conditions of Theorem 22.5. Thus, the sum is either zero or transcendental. However, as noted earlier, the former case is ruled out by the theorem of Baker, Birch and Wirsing. Thus, it is transcendental. By the previous theorem, the sum is equal to

$$\frac{1}{q} \left[\frac{\Gamma'}{\Gamma}(b/q) - \frac{\Gamma'}{\Gamma}(a/q) \right].$$

In this way, we see that the difference of any two values in the set

$$\frac{\Gamma'}{\Gamma}(a/q), \quad (a, q) = 1,$$

is transcendental. Thus, if there were at least two algebraic numbers in this set, we derive a contradiction. $\square$

Presumably, all the numbers in the set are transcendental. However, one is unable to establish this at present. Using Theorem 22.4, we can "solve" for $\Gamma'(a/q)/\Gamma(a/q)$ using the orthogonality relations for Dirichlet characters. To this end, we must first evaluate the sum

$$S_q := \sum_{(a,q)=1} \frac{\Gamma'}{\Gamma}(a/q).$$

We use the identity

$$\Gamma(z)\Gamma(z+1/q)\cdots\Gamma(z+(q-1)/q) = q^{1/2-qz}(2\pi)^{(q-1)/2}\Gamma(qz).$$

Logarithmically differentiating this and setting $z = 1/q$, we get

$$\sum_{a=1}^{q} \frac{\Gamma'}{\Gamma}(a/q) = -q\log q - \gamma q,$$

where we have used the fact that $\Gamma'(1) = -\gamma$. Thus,

$$\sum_{d|q} S_{q/d} = -q\log q - \gamma q$$

and we may apply Möbius inversion to solve for S_q:

$$S_q = -\sum_{d|q} \mu(d)\left(\frac{q}{d}\log\frac{q}{d} + \gamma\frac{q}{d}\right) = -\gamma\phi(q) - \sum_{d|q} \mu(d)\frac{q}{d}\log\frac{q}{d}.$$

We are now ready to prove:

Theorem 22.7 *For $(a,q) = 1$, we have*

$$-\frac{\phi(q)}{q}\frac{\Gamma'}{\Gamma}(a/q) = -\frac{S_q}{q} + \sum_{\chi\neq\chi_0} \overline{\chi}(a)L(1,\chi).$$

Proof. This is immediate from the orthogonality relations and our evaluation of S_q. $\square$

The interesting aspect of this formula is that it can be re-written as follows:

$$-\frac{\Gamma'}{\Gamma}(a/q) = \gamma + \frac{q}{\phi(q)}\sum_{d|q} \frac{\mu(d)}{d}\log\frac{q}{d} + \frac{q}{\phi(q)}\sum_{\chi\neq\chi_0} \overline{\chi}(a)L(1,\chi).$$

Apart from the γ on the right-hand side, we have a linear form in logarithms with algebraic coefficients and thus we immediately deduce:

Theorem 22.8 *For all $q > 1$ and $(a,q) = 1$, the number*

$$\frac{\Gamma'}{\Gamma}(a/q) + \gamma$$

is transcendental.

Proof. Baker's theorem tells us that it is either zero or transcendental. The former is not possible since $\Gamma'(x)/\Gamma(x)$ is a strictly increasing function for $x > 0$ (see Exercise 5 below). □

Exercises

1. Prove that

$$\zeta(k, a/q) + (-1)^k \zeta(k, 1 - a/q) = \frac{(-1)^{k-1}}{(k-1)!} \frac{d^{k-1}}{dz^{k-1}} (\pi \cot \pi z)|_{z=a/q}.$$

2. Show that $\frac{d^{k-1}}{dz^{k-1}}(\pi \cot \pi z)$ is π^k times a rational linear combination of expressions of the $\cot^r \pi z \csc^{2s} \pi z$ where $r + 2s = k$.

3. Conclude from the previous exercises that

$$\zeta(k, a/q) + (-1)^k \zeta(k, 1 - a/q) = i^k \pi^k \alpha_a$$

where α_a is an element in the qth cyclotomic field $\mathbb{Q}(\zeta_q)$.

4. Show that for an odd integer $k > 1$, $\zeta(k, 1/4)$ and $\zeta(k, 3/4)$ are linearly independent over $\mathbb{Q}$ if and only if $\zeta(k)/\pi^k$ is irrational.

5. Show that $\Gamma'(x)/\Gamma(x)$ is a strictly increasing function for $x > 0$.

Proof.

Exercises

Chapter 23

The Baker–Birch–Wirsing Theorem

We now give a detailed proof of the theorem of Baker, Birch and Wirsing introduced in the previous chapter. We present a somewhat modified version of their original proof by exploiting the properties of Dedekind determinants. These determinants have remarkable applications in a number of contexts in transcendence theory.

Theorem 23.1 *Let G be any finite abelian group of order n and $F : G \to \mathbb{C}$ be any complex valued function on G. The determinant of the $n \times n$ matrix given by $\left(F(xy^{-1})\right)$ as x, y range over the group elements is called the Dedekind determinant of F and is equal to*

$$\prod_{\chi} \left(\sum_{x \in G} \chi(x) F(x) \right),$$

where the product is over all characters χ of G.

Proof. Let V be the set of all functions from G to $\mathbb{C}$. This is an n-dimensional Hilbert space over $\mathbb{C}$ with an inner product

$$< g, h > = \frac{1}{n} \sum_{x \in G} g(x) \overline{h(x)}.$$

Let $\hat{G}$ be the set of all characters of G. These form an orthonormal basis for V. Now consider the linear map $T : V \to V$ whose values on a character χ are given by

$$T(\chi) = \left[\sum_{x \in G} \chi(x) F(x) \right] \chi.$$

M.R. Murty and P. Rath, *Transcendental Numbers*, DOI 10.1007/978-1-4939-0832-5_23, 131
© Springer Science+Business Media New York 2014

Clearly, the characters of G are eigenvectors of T and determinant of T is equal to

$$\prod_{\chi \in \hat{G}} \left(\sum_{x \in G} \chi(x) F(x) \right).$$

Now for every $x \in G$, let δ_x be the characteristic function of the set $\{x\}$. Then the set of all δ_x as x runs through elements of G also forms an orthogonal basis for V. We note that $< \delta_x, \chi >= \frac{1}{n}\chi(x^{-1})$ for any character χ of G and hence

$$\delta_x = \frac{1}{n} \sum_{\chi \in \hat{G}} \chi(x^{-1})\chi.$$

We have

$$
\begin{aligned}
T(\delta_x) &= \frac{1}{n} \sum_{\chi \in \hat{G}} \chi(x^{-1}) T(\chi) \\
&= \frac{1}{n} \sum_{\chi \in \hat{G}} \chi(x^{-1}) \left[\sum_{y \in G} \chi(y) F(y) \right] \chi \\
&= \frac{1}{n} \sum_{\chi \in \hat{G}} \sum_{y \in G} \chi(x^{-1}y) F(y) \chi \\
&= \frac{1}{n} \sum_{\chi \in \hat{G}} \sum_{z \in G} \chi(z^{-1}) F(xz^{-1}) \chi \\
&= \sum_{z \in G} F(xz^{-1}) \frac{1}{n} \sum_{\chi \in \hat{G}} \chi(z^{-1}) \chi = \sum_{z \in G} F(xz^{-1}) \delta_z.
\end{aligned}
$$

Thus the matrix $\left(F(xy^{-1}) \right)_{x,y \in G}$ is simply the matrix of T with respect to the basis $\{\delta_x : x \in G \}$. This proves the theorem. $\square$

Let us now prove the Baker–Birch–Wirsing theorem. We are given a non-zero periodic arithmetic function f with period q. Further, f takes algebraic values and $f(n) = 0$ whenever $1 < (n, q) < q$. Finally, we are given that the q-th cyclotomic polynomial is irreducible over $\mathbb{Q}(f(1), \ldots, f(q))$. We need to show that

$$\sum_{n=1}^{\infty} \frac{f(n)}{n} \neq 0.$$

Recall that the digamma function $\psi(z)$ for $z \neq -n$ with $n \in \mathbb{N}$ is the logarithmic derivative $\frac{\Gamma'(z)}{\Gamma(z)}$ of the Γ-function and is given by

$$-\psi(z) = \gamma + \frac{1}{z} + \sum_{n \geq 1} \left(\frac{1}{n+z} - \frac{1}{n} \right).$$

As shown in the previous chapter, the series converges if and only if

$$\sum_{a=1}^{q} f(a) = 0$$

and in which case, we note that

$$L(1, f) = \sum_{n=1}^{\infty} \frac{f(n)}{n} = \frac{-1}{q} \sum_{(a,q)=1} f(a)\psi(a/q) - \frac{f(q)\psi(1)}{q}.$$

Since $\psi(1) = -\gamma$ and $\sum_{a=1}^{q} f(a) = 0$, we have

$$f(q) = -\sum_{(a,q)=1} f(a)$$

so that

$$L(1, f) = \frac{-1}{q} \sum_{(a,q)=1} f(a)(\psi(a/q) + \gamma).$$

Also

$$L(1, f) = -\sum_{a=1}^{q-1} \hat{f}(a) \log(1 - \zeta_q^a),$$

where $\hat{f}$ is the Fourier transform of f and $\zeta_q = e^{2\pi i/q}$. Let F be the field $\mathbb{Q}(f(1), \ldots, f(q))$ and

$$\log(1 - \zeta_q^{\alpha_1}), \ldots, \log(1 - \zeta_q^{\alpha_t})$$

be a maximal F-linear independent subset of

$$\{\log(1 - \zeta_q^a) \mid 1 \leq a \leq q - 1\}.$$

Then

$$\log(1 - \zeta_q^a) = \sum_{b=1}^{t} A_{ab} \log(1 - \zeta_q^{\alpha_b}),$$

where $A_{ab} \in F$. Then by the given hypothesis, we have

$$\beta_1 \log(1 - \zeta_q^{\alpha_1}) + \cdots + \beta_t \log(1 - \zeta_q^{\alpha_t}) = 0$$

where

$$\beta_b = \sum_{a=1}^{q-1} \hat{f}(a) A_{ab}.$$

Since f takes values in F, $\hat{f}$ is algebraic valued. Thus by Baker's theorem on linear forms in logarithms, we have

$$\beta_b = \sum_{a=1}^{q-1} \hat{f}(a) A_{ab} = 0, \qquad 1 \leq b \leq t.$$

Then for any automorphism $\sigma \in \mathrm{Gal}(F(\zeta_q)/F)$, we have

$$\sum_{a=1}^{q-1} \sigma(\hat{f}(a))\mathrm{A}_{ab} = 0, \qquad 1 \leq b \leq t,$$

and hence

$$\sum_{a=1}^{q-1} \sigma(\hat{f}(a))\log(1 - \zeta_q^a) = 0.$$

Let G be the Galois group of the extension $F(\zeta_q)/F$. We note that G is isomorphic to the group $(\mathbb{Z}/q\mathbb{Z})^*$. For $(h,q) = 1$, let $\sigma_h \in G$ be such that

$$\sigma_h(\zeta_q) = \zeta_q^h.$$

Define $f_h(n) := f(nh^{-1})$ for $(h,q) = 1$. Then, we have

$$\sum_{(a,q)=1} f_h(a) = -f_h(q) = -f(q) \quad \text{and} \quad \sigma_h(\hat{f}(n)) = \widehat{f_h}(n).$$

Hence

$$L(1, f_h) = \sum_{n=1}^{\infty} \frac{f_h(n)}{n} = -\sum_{a=1}^{q-1} \widehat{f_h}(a)\log(1 - \zeta_q^a)$$

$$= -\sum_{a=1}^{q-1} \sigma_h(\hat{f}(a))\log(1 - \zeta_q^a) = 0$$

for all $(h,q) = 1$. This gives that

$$L(1, f_h) = \frac{-1}{q} \sum_{(a,q)=1} f_h(a)(\psi(a/q) + \gamma) = 0.$$

Hence by making a change of variable, we have

$$\sum_{(a,q)=1} f(a)(\psi(ah/q) + \gamma) = 0, \tag{23.1}$$

where it is implied that ah is taken to be the reduced residue class $b \pmod{q}$ satisfying

$$ah \equiv b \pmod{q}.$$

Now

$$\mathrm{A} := (\psi(ah/q) + \gamma)_{(ah,q)=1}$$

is a Dedekind matrix on the group $\mathrm{H} = (\mathbb{Z}/q\mathbb{Z})^*$ and its determinant (up to a sign) is given by

$$\prod_{\chi \in \hat{\mathrm{H}}} \left(\sum_{h \in \mathrm{H}} \chi(h)(\psi(h/q) + \gamma) \right).$$

If we show that the matrix A is invertible, then f vanishes everywhere and we are done. For a non-principal character χ of H,

$$\sum_{h \in H} \chi(h)\,(\psi(h/q) + \gamma) = -q\,L(1, \chi).$$

It is classical that $L(1, \chi) \neq 0$ for $\chi \neq 1$. Thus we need only to verify that

$$\sum_{h \in H} (\psi(h/q) + \gamma) \neq 0. \tag{23.2}$$

Since $\psi(x)$ is an increasing function and $\psi(1) = -\gamma$, we have the above identity. This completes the proof of the theorem.

In a recent work [63], a generalisation of the above theorem has been derived.

Exercises

1. Let q be an odd prime. Using the fact that the numbers

$$\eta_a := \frac{\sin \pi a/q}{\sin \pi/q}, \qquad 1 < a < q/2$$

   are multiplicatively independent units in the cyclotomic field, $\mathbb{Q}(e^{2\pi i/q})$, show that the numbers

$$\log \eta_a, \qquad 1 < a < q/2$$

   are linearly independent over the field of algebraic numbers.

2. Apply the previous exercise to show that if f is a rational valued even periodic function with a prime period q, then

$$\sum_{n=1}^{\infty} \frac{f(n)}{n} \neq 0.$$

3. If f as an odd rational-valued function periodic function with a prime period q, then show that

$$\sum_{n=1}^{\infty} \frac{f(n)}{n} \neq 0.$$

   In fact, when the sum converges, show that it is an algebraic multiple of π.

4. Show that the two conditions in the statement of the Baker–Birch–Wirsing theorem, namely $f(n) = 0$ whenever $1 < (n, q) < q$ and the q-th cyclotomic polynomial is irreducible over $\mathbb{Q}(f(1), \ldots, f(q))$, are both necessary.

Chapter 24

Transcendence of Some Infinite Series

In this chapter, we investigate the transcendental nature of the sum

$$\sideset{}{'}\sum_{n \in \mathbb{Z}} \frac{A(n)}{B(n)}$$

where $A(x), B(x)$ are polynomials with algebraic coefficients with $\deg A < \deg B$ and the sum is over integers n which are not zeros of $B(x)$. We relate this question to a conjecture originally due to Schneider. A stronger version of this conjecture was later suggested by Gel'fond and Schneider. In certain cases, these conjectures are known and this allows one to obtain some unconditional results of a general nature.

Let $A(x)$ and $B(x)$ be polynomials in $\overline{\mathbb{Q}}[x]$ with $\deg A < \deg B$ so that $B(x)$ has no integral zeros. We will evaluate the infinite series

$$\sum_{n \in \mathbb{Z}} \frac{A(n)}{B(n)}, \tag{24.1}$$

interpreted as

$$\lim_{N \to \infty} \sum_{|n| \leq N} \frac{A(n)}{B(n)}.$$

We seek to determine under what conditions the sum is a transcendental number. One could also allow $B(x)$ to have integral zeros and exclude these integral zeros from the sum (24.1). The methods described in this chapter apply to this general setting also. We will follow the treatment given by Murty and Weatherby in [100].

M.R. Murty and P. Rath, *Transcendental Numbers*, DOI 10.1007/978-1-4939-0832-5_24, 137
© Springer Science+Business Media New York 2014

In 1934, Gel'fond [47] and Schneider [110] independently solved Hilbert's seventh problem which predicted the following result and which we have seen before: if α is an algebraic number $\neq 0, 1$ and β is an irrational algebraic number, then α^β is transcendental. This result has some interesting consequences. For example, by taking $\alpha = -1$ and $\beta = -i = -\sqrt{-1}$, we deduce the transcendence of e^π. Similarly, one can deduce the transcendence of $e^{\pi\beta}$ for any real algebraic number β. Based on their investigations, Gel'fond and Schneider were led to formulate some general conjectures that provided a concrete goal for researchers in subsequent decades. Let us begin with the following conjecture of Schneider:

If $\alpha \neq 0, 1$ is algebraic and β is an algebraic irrational of degree $d \geq 2$, then

$$\alpha^\beta, \ldots, \alpha^{\beta^{d-1}}$$

are algebraically independent.

In 1949, Gel'fond [48] proved that if $d \geq 3$, then the transcendence degree of

$$\mathbb{Q}(\alpha^\beta, \ldots, \alpha^{\beta^{d-1}})$$

is at least 2. Thus, in the case $d = 3$, this proves Schneider's conjecture. Building on earlier work of Chudnovsky [33] and Philippon [92], Diaz [43] showed that

$$\text{tr.deg.} \, \mathbb{Q}(\alpha^\beta, \ldots, \alpha^{\beta^{d-1}}) \geq \left[\frac{d+1}{2}\right].$$

Thus, we have crossed the "midway" point in our journey towards Schneider's conjecture.

Shortly after their solution to Hilbert's seventh problem, Gel'fond and Schneider were led to formulate a more general conjecture:

If α is algebraic and unequal to $0, 1$, and β is algebraic of degree $d \geq 2$, then

$$\log \alpha, \alpha^\beta, \ldots, \alpha^{\beta^{d-1}}$$

are algebraically independent.

We refer to this assertion as the Gel'fond–Schneider conjecture. We point out that, as will be seen, for our purposes α is a root of unity and we use these conjectures only in a special case.

In the study of the transcendence properties of the series (24.1), the case where the roots of $B(x)$ are rational and non-integral is easy. As will be evident from the discussion below, the sum in this case is equal to $\pi P(\pi)$, where $P(x) \in \overline{\mathbb{Q}}[x]$. Thus, if all the roots are rational and non-integral, the sum (24.1) is either zero or transcendental. We seek to establish a similar theorem in the general case when $B(x)$ has irrational roots.

One of the main theorems of this chapter is the following:

Theorem 24.1 *Let* $A(x), B(x) \in \overline{\mathbb{Q}}[x]$ *with* $\deg A < \deg B$, $A(x)$ *coprime to* $B(x)$, *and* $A(x)$ *not identically zero. Suppose that the roots of* $B(x)$ *are given by* $-r_1, \ldots, -r_l \in \mathbb{Q} \setminus \mathbb{Z}$ *and* $-\alpha_1, \ldots, -\alpha_k \notin \mathbb{Q}$ *so that all roots are simple and* $\alpha_i \pm \alpha_j \notin \mathbb{Q}$ *for* $i \neq j$. *If* $k = 0$, *then the series*

$$S = \sum_{n \in \mathbb{Z}} \frac{A(n)}{B(n)}$$

is an algebraic multiple of π. *If* $k \geq 1$, *then Schneider's conjecture implies that* S/π *is transcendental and the Gel'fond–Schneider conjecture implies that* S *and* π *are algebraically independent.*

Proof. We begin the proof with the following two observations. The first is that

$$\pi \cot \pi x = \sum_{n \in \mathbb{Z}} \frac{1}{n + x},$$

which is valid for $x \notin \mathbb{Z}$. Now,

$$\cot \pi x = i \frac{e^{i\pi x} + e^{-i\pi x}}{e^{i\pi x} - e^{-i\pi x}} = i \frac{e^{2\pi i x} + 1}{e^{2\pi i x} - 1} = i + \frac{2i}{e^{2\pi i x} - 1},$$

and this will be useful below. The second is that by the theory of partial fractions, we can write

$$\frac{A(x)}{B(x)} = \sum_{m=1}^{l} \frac{d_m}{x + r_m} + \sum_{j=1}^{k} \frac{c_j}{x + \alpha_j}.$$

By direct calculation, our series divided by π is equal to

$$i \left(\sum_{j=1}^{k} c_j \frac{e^{2\pi i \alpha_j} + 1}{e^{2\pi i \alpha_j} - 1} + \sum_{m=1}^{l} d_m \frac{e^{2\pi i r_m} + 1}{e^{2\pi i r_m} - 1} \right),$$

where each c_j and d_m is in $\overline{\mathbb{Q}} \setminus \{0\}$. If all of the roots are rational, the first sum is empty and S/π is algebraic which proves the first assertion.

Assume that $B(x)$ has at least one irrational root and suppose that the sum S is an algebraic multiple of π. We have

$$S/\pi - i \sum_{m=1}^{l} d_m \frac{e^{2\pi i r_m} + 1}{e^{2\pi i r_m} - 1} = i \sum_{j=1}^{k} c_j \frac{e^{2\pi i \alpha_j} + 1}{e^{2\pi i \alpha_j} - 1}$$

$$= i \sum_{j=1}^{k} c_j + 2i \sum_{j=1}^{k} \frac{c_j}{e^{2\pi i \alpha_j} - 1}$$

so that

$$\sum_{j=1}^{k} \frac{c_j}{e^{2\pi i \alpha_j} - 1} = \frac{1}{2i}\left(S/\pi - i\sum_{m=1}^{l} d_m \frac{e^{2\pi i r_m}+1}{e^{2\pi i r_m}-1} - i\sum_{j=1}^{k} c_j\right) = \theta \in \overline{\mathbb{Q}}.$$

By assumption, $[\mathbb{Q}(\alpha_1,\ldots,\alpha_k):\mathbb{Q}] = d > 1$. Now by the theorem of the primitive element, there is a $\beta \in \overline{\mathbb{Q}}$ of degree d such that $\mathbb{Q}(\alpha_1,\ldots,\alpha_k) = \mathbb{Q}(\beta)$. Thus, we have the equations

$$\alpha_j = \sum_{a=0}^{d-1} r_{a,j}\beta^a$$

where each $r_{a,j} \in \mathbb{Q}$. Let us chose an integer $M \in \mathbb{Z}$ such that

$$\alpha_j = \frac{1}{M}\sum_{a=0}^{d-1} n_{a,j}\beta^a$$

where each $n_{a,j} \in \mathbb{Z}$. Let $\alpha = e^{\pi i/M}$. If Schneider's conjecture is true, then the numbers

$$\alpha^\beta, \ldots, \alpha^{\beta^{d-1}}$$

are algebraically independent and hence

$$\alpha^{2\beta}, \ldots, \alpha^{2\beta^{d-1}}$$

are also algebraically independent. Define $x_a := \alpha^{2\beta^a} = e^{2\pi i \beta^a/M}$ for $a = 1,\ldots,d-1$ so that

$$e^{2\pi i \alpha_j} = e^{\frac{2\pi i}{M}\sum_{a=0}^{d-1} n_{a,j}\beta^a} = \gamma_j x_1^{n_{1,j}}\cdots x_{d-1}^{n_{d-1,j}}$$

where $\gamma_j = e^{2\pi i n_{0,j}/M}$ is a root of unity.

Making this substitution, we have

$$\theta = \sum_{j=1}^{k} \frac{c_j}{\gamma_j x_1^{n_{1,j}}\cdots x_{d-1}^{n_{d-1,j}} - 1}.$$

This implies that all of the x_i's cancel in some fashion leaving only an algebraic number. We will now show that this does not occur under the conditions of our theorem.

Let us examine the function

$$F(X_1,\ldots,X_{d-1}) = \sum_{j=1}^{k} \frac{c_j}{\gamma_j X_1^{n_{1,j}}\cdots X_{d-1}^{n_{d-1,j}} - 1}.$$

If we can show that F is not constant, then our sum actually contains some variables and we are done. We show that F is not constant by examining F at some special points. Let y be a new indeterminate. For some integral values $e_1, \ldots, e_{d-1}$ to be specified later, let $X_i = y^{e_i}$. We have that

$$F(y^{e_1}, \ldots, y^{e_{d-1}}) = \sum_{j=1}^{k} \frac{c_j}{\gamma_j y^{\overline{n_j} \cdot \overline{e}} - 1}$$

where $\overline{n_j} = (n_{1,j}, \ldots, n_{d-1,j})$ and $\overline{e} = (e_1, \ldots, e_{d-1})$. For any $\overline{n_j} \cdot \overline{e} < 0$, we have

$$\frac{1}{\gamma_j y^{\overline{n_j} \cdot \overline{e}} - 1} = -1 - \frac{1}{\gamma_j^{-1} y^{-\overline{n_j} \cdot \overline{e}} - 1}$$

so that

$$F(y^{e_1}, \ldots, y^{e_{d-1}}) = - \sum_{\overline{n_j} \cdot \overline{e} < 0} c_j + \sum_{j=1}^{k} c_j \frac{\operatorname{sgn}(\overline{n_j} \cdot \overline{e})}{\gamma_j^{\operatorname{sgn}(\overline{n_j} \cdot \overline{e})} y^{|\overline{n_j} \cdot \overline{e}|} - 1} \qquad (24.2)$$

where $\operatorname{sgn}(x) = 1$ if $x \geq 0$ and -1 otherwise. If every power of y that appears in the second sum is different and non-zero, then we can group each summand over a common denominator and notice that the degree of the numerator will be less than the degree of the denominator. It is easy to see that if the function above in (24.2) (as a function of y) is constant, then each $c_j = 0$, which is a contradiction. Hence, if we can guarantee the condition that each $|\overline{n_j} \cdot \overline{e}|$ is different and non-zero, then our function is not constant, and therefore the transcendental part of our original series does not vanish and we are done.

We now specify $\overline{e}$. We wish to choose integers e_i such that $\overline{n_j} \cdot \overline{e} \neq \pm \overline{n_{j'}} \cdot \overline{e}$ for $j \neq j'$. We also need each $\overline{n_j} \cdot \overline{e} \neq 0$ as well. Thus, we need $\overline{e}$ which simultaneously satisfies

$$(\overline{n_j} \pm \overline{n_{j'}}) \cdot \overline{e} \neq 0$$
$$\overline{n_j} \cdot \overline{e} \neq 0.$$

To find such an $\overline{e}$, we use a lattice point argument. For positive integer D, let $I_D = (0, D]$. Examine the box $B_D = I_D^{d-1}$ which contains a total of D^{d-1} lattice points. We wish to avoid points which satisfy the equations

$$(\overline{n_j} \pm \overline{n_{j'}}) \cdot \overline{e} = 0$$
$$\overline{n_j} \cdot \overline{e} = 0.$$

Our conditions on the irrational roots ensure that $\overline{n_j} \pm \overline{n_{j'}} \neq \overline{0}$ so that none of these equations is trivially satisfied. There are at most D^{d-2} lattice points in B_D which satisfy each equation. We have $2\binom{k}{2}$ equations of the first form and k equations of the second type. Thus for D large enough, we have at least

$$D^{d-1} - \left(k + 2\binom{k}{2}\right) D^{d-2} > 1$$

lattice points to choose from for $\bar{e}$. Thus, there exists such an $\bar{e}$ which shows that our function F is not constant. This shows that θ, and therefore S/π is transcendental and we have the second assertion of our theorem. To show the third assertion, we observe that the Gelfond–Schneider conjecture predicts that the d numbers

$$\log(\alpha), \alpha^\beta, \ldots, \alpha^{\beta^{d-1}}$$

are algebraically independent. In our setting, this conjecture implies that π and $x_1, \ldots, x_{d-1}$ are algebraically independent which completes the argument. $\square$

The condition that $B(x)$ has only non-integral roots is not a serious constraint. In fact, it can easily be removed in some cases if we understand that we are considering sums

$$\sideset{}{'}\sum_{n\in\mathbb{Z}} \frac{A(n)}{B(n)} \qquad (24.3)$$

where the dash on the sum means that we sum over only those integers n which are not roots of $B(x)$. More precisely, we now indicate how Theorem 24.1 is valid (partially) if we remove the restriction that $B(x)$ has no integral roots and we interpret the sum (24.3) as omitting the integral zeros of $B(x)$. Indeed, suppose that $-n_1, \ldots, -n_t$ are all the integral roots of $B(x)$. After expanding $A(x)/B(x)$ in partial fractions, we encounter three types of sums:

$$\sideset{}{'}\sum_{n\in\mathbb{Z}} \frac{1}{n+n_i}, \quad \sideset{}{'}\sum_{n\in\mathbb{Z}} \frac{1}{n+r_i} \quad \text{and} \quad \sideset{}{'}\sum_{n\in\mathbb{Z}} \frac{1}{n+\alpha_i}. \qquad (24.4)$$

Clearly, in relation to transcendence, the first sum above has no effect.

The second and third sums of (24.4) are

$$\pi \cot \pi r_i - \sum_{j=1}^{t} \frac{1}{n_j+r_i} \quad \text{and} \quad \pi \cot \pi \alpha_i - \sum_{j=1}^{t} \frac{1}{n_j+\alpha_i}.$$

Since the second sum for each is algebraic, it is clear that when $B(x)$ has at least one integral zero we will obtain a similar conclusion to the last part of Theorem 24.1 where there are no integral zeroes. More precisely, in the same setting of Theorem 24.1 with $k \geq 1$, allowing $B(x)$ to possibly have integral roots, the Gelfond–Schneider conjecture implies that the sum (24.3) and π are algebraically independent.

Since the Gelfond–Schneider conjecture is still far away from being established, and we are somewhat "nearer" to the Schneider conjecture, it is reasonable to ask what can be said about the number S in the previous theorem assuming this weaker conjecture. Here one has the following.

Theorem 24.2 *Fix nonconstant polynomials $A_1(x), A_2(x), B_1(x), B_2(x) \in \overline{\mathbb{Q}}[x]$ so that $A_i(x)$ has no common factors with $B_i(x)$, $\deg(A_i) < \deg(B_i)$ and the functions $A_1(x)/B_1(x), A_2(x)/B_2(x)$ are not scalar multiples. Write*

$B(x) = \mathrm{lcm}(B_1(x), B_2(x))$ *and suppose that* $B(x)$ *has only simple irrational roots given by* $-\alpha_1, \ldots, -\alpha_k$ *such that* $\alpha_i \pm \alpha_j \notin \mathbb{Q}$ *for* $i \neq j$. *If Schneider's conjecture is true, then the quotient*

$$\left(\sum_{n \in \mathbb{Z}} \frac{A_1(n)}{B_1(n)} \right) \Big/ \left(\sum_{n \in \mathbb{Z}} \frac{A_2(n)}{B_2(n)} \right)$$

is transcendental.

Proof. We first work with the case that $B_1(x)$ and $B_2(x)$ are scalar multiples. Without loss of generality, we can assume that $B_1(x) = B_2(x) = B(x)$. By partial fractions we write

$$\frac{A_1(x)}{B(x)} = \sum_{j=1}^{k} \frac{c_j}{x + \alpha_j}$$

and

$$\frac{A_2(x)}{B(x)} = \sum_{j=1}^{k} \frac{C_j}{x + \alpha_j}$$

for some $c_j, C_j \in \overline{\mathbb{Q}}$. As in the proof of Theorem 24.1 , we have

$$\sum_{n \in \mathbb{Z}} \frac{A_1(n)}{B(n)} = \pi i (\beta_1 + 2\theta_1), \quad \sum_{n \in \mathbb{Z}} \frac{A_2(n)}{B(n)} = \pi i (\beta_2 + 2\theta_2)$$

where

$$\beta_1 = \sum_{j=1}^{k} c_j, \quad \beta_2 = \sum_{j=1}^{k} C_j, \quad \theta_1 = \sum_{j=1}^{k} \frac{c_j}{e^{2\pi i \alpha_j} - 1}, \quad \theta_2 = \sum_{j=1}^{k} \frac{C_j}{e^{2\pi i \alpha_j} - 1}.$$

Theorem 24.1 implies that θ_1 and θ_2 are transcendental. If the ratio of the two series is algebraic then

$$\sum_{n \in \mathbb{Z}} \frac{A_1(n)}{B(n)} - \lambda \sum_{n \in \mathbb{Z}} \frac{A_2(n)}{B(n)} = 0$$

for some algebraic $\lambda \neq 0$. Thus

$$2(\theta_1 - \lambda\theta_2) = \lambda\beta_2 - \beta_1.$$

We now focus on

$$\theta_1 - \lambda\theta_2 = \sum_{j=1}^{k} \frac{c_j - \lambda C_j}{e^{2\pi i \alpha_j} - 1}.$$

Similar to the proof of Theorem 24.1, we see that $\theta_1 - \lambda\theta_2$ is algebraic only if $c_j - \lambda C_j = 0$ for each j. This implies that $A_1(x) = \lambda A_2(x)$ which gives a contradiction.

Next we assume that $B_1(x) \neq \alpha B_2(x)$ for any algebraic number α. That is, without loss of generality, $B_2(x)$ has a root R such that $B_1(R) \neq 0$. Suppose that the quotient

$$\left(\sum_{n \in \mathbb{Z}} \frac{A_1(n)}{B_1(n)} \right) \bigg/ \left(\sum_{n \in \mathbb{Z}} \frac{A_2(n)}{B_2(n)} \right)$$

is algebraic. Inserting the appropriate missing factors to each numerator respectively, we have that the quotient

$$\left(\sum_{n \in \mathbb{Z}} \frac{\widetilde{A_1}(n)}{B(n)} \right) \bigg/ \left(\sum_{n \in \mathbb{Z}} \frac{\widetilde{A_2}(n)}{B(n)} \right)$$

is algebraic. We see that we are in a situation close to the previous case. We remark that in the previous case, $A_i(x)$ need not be coprime with $B_i(x) = B(x)$. If there were common factors, some of the (say) c_j's would simply be zero and we would still obtain the same contradiction. With this in mind, if the quotient of series is algebraic, then according to the previous case, there is a non-zero $\lambda \in \overline{\mathbb{Q}}$ such that

$$\frac{\widetilde{A_1}(x)}{B(x)} = \lambda \frac{\widetilde{A_2}(x)}{B(x)}$$

which simplifies to

$$\frac{A_1(x)}{B_1(x)} = \lambda \frac{A_2(x)}{B_2(x)}.$$

Since R is a pole of the right side but not the left, we have a contradiction and we are done. $\square$

A simple corollary of Theorem 24.2 is that by assuming Schneider's conjecture, along with our condition on the irrational roots of $B(x)$, one can establish the transcendence of S with "at most one exception".

Both the Schneider conjecture and the Gelfond–Schneider conjecture are special cases of the far-reaching Schanuel's conjecture. As discussed earlier, this conjecture predicts that if the complex numbers $x_1, \ldots, x_n$ are linearly independent over $\mathbb{Q}$, then

$$\text{tr.deg.} \, \mathbb{Q}(x_1, \ldots, x_n, e^{x_1}, \ldots, e^{x_n}) \geq n.$$

An interesting consequence of this conjecture is that π and e are algebraically independent. If $x_1, \ldots, x_n$ are algebraic numbers, the assertion of the Schanuel's conjecture is the Lindemann–Weierstrass theorem. Beyond this, the general conjecture seems unreachable at present. However, as mentioned in the introduction, progress has been made on the Schneider conjecture and this allows one to make a some of these results unconditional. To standardise the setting throughout, let $K_1 = \mathbb{Q}(\alpha_1, \ldots, \alpha_k)$, the field generated by the roots of $B(x)$, and let K_2 be K_1 adjoin the coefficients of $A(x)$ and $B(x)$. Restricting ourselves to the case where $B(x)$ has simple roots, the following are unconditional versions of Theorems 24.1 and 24.2, respectively.

Theorem 24.3 *In the same setting as Theorem 24.1, if* $[K_1 : \mathbb{Q}] = 2$ *or* 3, *then* S/π *is transcendental. If* K_1 *is an imaginary quadratic field, then* S *is algebraically independent from* π.

Theorem 24.4 *In the same setting as Theorem 24.2, if* $[K_1 : \mathbb{Q}] = 2$ *or* 3, *then the quotient*

$$\left(\sum_{n \in \mathbb{Z}} \frac{A_1(n)}{B_1(n)} \right) \bigg/ \left(\sum_{n \in \mathbb{Z}} \frac{A_2(n)}{B_2(n)} \right)$$

is transcendental.

Here is the proof of the above theorems. Since Schneider's conjecture is true for $d = 2, 3$ (Gel'fond), we immediately have Theorem 24.4 and the first part of Theorem 24.3. To prove the second part of Theorem 24.3, we invoke the theorem of Nesterenko [87] discussed before, namely, if $\mathbb{Q}(\sqrt{-D})$ is an imaginary quadratic field with $D > 0$, then π and $e^{\pi\sqrt{D}}$ are algebraically independent. Thus, S is algebraically independent from π.

In the above theorems, we restricted ourselves to the case of simple roots. We can also derive results in the case of multiple roots. For this, we shall need the following lemma regarding derivatives of the cotangent function.

Lemma 24.5 *For* $k \geq 2$ *and* $x \notin \mathbb{Z}$,

$$\frac{d^{k-1}}{dx^{k-1}}(\pi \cot(\pi x)) = (2\pi i)^k \left(\frac{A_{k,1}}{e^{2\pi i x} - 1} + \cdots + \frac{A_{k,k}}{(e^{2\pi i x} - 1)^k} \right)$$

where each $A_{i,j} \in \mathbb{Z}$ *with* $A_{k,1}, A_{k,k} \neq 0$.

Proof. We have that

$$\pi \cot(\pi x) = \pi i + 2\pi i/(e^{2\pi i x} - 1).$$

Differentiating this we obtain the result for $k = 2$. Assuming that the equality is true for all $k < t$. Then by induction we have $A_{t-1,1}, \ldots, A_{t-1,t-1} \in \mathbb{Z}$ with $A_{t-1,1}, A_{t-1,t-1} \neq 0$ such that

$$\frac{d}{dx} \left(\frac{d^{t-2}}{dx^{t-2}}(\pi \cot(\pi x)) \right) = (2\pi i)^{t-1} \frac{d}{dx} \left(\frac{A_{t-1,1}}{e^{2\pi i x} - 1} + \cdots + \frac{A_{t-1,t-1}}{(e^{2\pi i x} - 1)^{t-1}} \right).$$

This is equal to

$$(2\pi i)^t \left(-A_{t-1,1} \frac{e^{2\pi i x}}{(e^{2\pi i x} - 1)^2} - \cdots - (t-1)A_{t-1,t-1} \frac{e^{2\pi i x}}{(e^{2\pi i x} - 1)^t} \right).$$

By subtracting and adding 1 from each numerator, we have

$$(2\pi i)^t \left(-A_{t-1,1} \frac{e^{2\pi i x} - 1 + 1}{(e^{2\pi i x} - 1)^2} - \cdots - (t-1)A_{t-1,t-1} \frac{e^{2\pi i x} - 1 + 1}{(e^{2\pi i x} - 1)^t} \right)$$

which equals $(2\pi i)^t$ times

$$\left(-\frac{A_{t-1,1}}{e^{2\pi ix} - 1} - \frac{A_{t-1,1}}{(e^{2\pi ix} - 1)^2} - \cdots - \frac{(t-1)A_{t-1,t-1}}{(e^{2\pi ix} - 1)^{t-1}} - \frac{(t-1)A_{t-1,t-1}}{(e^{2\pi ix} - 1)^t}\right)$$

and this gives the result. $\square$

Since

$$\sum_{n\in\mathbb{Z}} \frac{1}{n + x} = \pi\cot(\pi x) = \pi i + \frac{2\pi i}{e^{2\pi ix} - 1},$$

a consequence of Lemma 24.5 is that for each $k \geq 2$,

$$\sum_{n\in\mathbb{Z}} \frac{1}{(n + x)^k} = \frac{(-1)^{k-1}(2\pi i)^k}{(k-1)!}\left(\frac{A_{k,1}}{e^{2\pi ix} - 1} + \cdots + \frac{A_{k,k}}{(e^{2\pi ix} - 1)^k}\right) \qquad (24.5)$$

for $A_{k,j}$'s as above.

We are now ready to consider the case of multiple roots. Let us start with the following theorem.

Theorem 24.6 *If the Gelfond–Schneider conjecture is true, then for any $A(x), B(x)$ lying in $\overline{\mathbb{Q}}[x]$ with $\deg(A) < \deg(B)$ and $B(n) \neq 0$ for any $n \in \mathbb{Z}$, the series*

$$\sum_{n\in\mathbb{Z}} \frac{A(n)}{B(n)}$$

is either zero or transcendental.

Before we start the proof of the above theorem, it is useful to remark that if $B(x)$ has only rational (and not integral) roots, then it is not hard to see from the previous lemma that the value of (24.1) is a polynomial in π with algebraic coefficients and zero constant term. Thus, again the sum is either zero or transcendental. So we can focus on the case of irrational roots. Indeed, if we also allow $-n_1, \ldots, -n_t$ to be integral roots and understand the sum over $\mathbb{Z}$ excludes these integral roots, we are led to study, as before, sums of three types:

$$\sideset{}{'}\sum_{n\in\mathbb{Z}} \frac{1}{(n + n_i)^k}, \quad \sideset{}{'}\sum_{n\in\mathbb{Z}} \frac{1}{(n + r_i)^k} \quad \text{and} \quad \sideset{}{'}\sum_{n\in\mathbb{Z}} \frac{1}{(n + \alpha_j)^k}. \qquad (24.6)$$

The third sum is

$$\frac{(-1)^{k-1}}{(k-1)!} D^{k-1}(\pi\cot\pi x)\Big|_{x=\alpha_i} - \sum_{j=1}^{t} \frac{1}{(n_j + \alpha_i)^k},$$

and the last sum is algebraic. A similar comment applies for the middle sum, which turns out to be an algebraic multiple of π^k plus a rational number. Finally, the first sum is easily seen to be a rational multiple of π^k plus a rational number. Thus, in the case that there are integral roots and we sum over those $n \in \mathbb{Z}$

which exclude those roots, we are able to assert the stronger theorem that the series is either given explicitly as an algebraic number, seen as the sum of the remainder terms $\sum_{j=1}^{t}$ above, or is transcendental under the assumption of the Gelfond–Schneider conjecture.

We can now proceed to prove Theorem 24.6. Let $-\alpha_1, \ldots, -\alpha_k \in \overline{\mathbb{Q}} \setminus \mathbb{Z}$ be the roots of $B(x)$ with multiplicities $m_1, \ldots, m_k$, respectively. By partial fractions we write

$$\frac{A(x)}{B(x)} = \sum_{j=1}^{k} \sum_{l=1}^{m_j} \frac{c_{j,l}}{(x+\alpha_j)^l}.$$

By Lemma 24.5 we have that $\sum_{n \in \mathbb{Z}} A(n)/B(n)$ is equal to

$$\pi i \sum_{j=1}^{k} c_{j,1} \frac{e^{2\pi i \alpha_j} + 1}{e^{2\pi i \alpha_j} - 1} + \tag{24.7}$$

$$\sum_{j=1}^{k} \sum_{l=2}^{m_j} \frac{c_{j,l}(-1)^{l-1}(2\pi i)^l}{(l-1)!} \left(\frac{A_{l,1}}{e^{2\pi i \alpha_j} - 1} + \cdots + \frac{A_{l,l}}{(e^{2\pi i \alpha_j} - 1)^l} \right).$$

Viewing this as a polynomial in π (with zero constant term), we analyse the coefficients. By the primitive element theorem, there is an algebraic β of degree d such that $\mathbb{Q}(\beta) = \mathbb{Q}(\alpha_1, \ldots, \alpha_k)$. Thus, as before, we can write each

$$\alpha_j = \frac{1}{M} \sum_{a=0}^{d-1} n_{a,j} \beta^a$$

for some integers $M, n_{a,j}$ so that

$$e^{2\pi i \alpha_j} = \prod_{a=0}^{d-1} e^{2\pi i n_{a,j} \beta^a / M}.$$

Let $\alpha = e^{\pi i / M}$ so that we have that each coefficient of a given power of π in (24.7) lies in the field $\overline{\mathbb{Q}}(\alpha^\beta, \ldots, \alpha^{\beta^{d-1}})$. Since the Gelfond–Schneider conjecture implies that $\pi, \alpha^\beta, \ldots, \alpha^{\beta^{d-1}}$ are algebraically independent, we conclude that the sum is either zero or transcendental. This completes the proof of Theorem 24.6.

One also has the following theorem in this context.

Theorem 24.7 Let $A(x), B(x) \in \overline{\mathbb{Q}}[x]$ with $\deg(A) < \deg(B)$, and $A(x)$ coprime to $B(x)$. Suppose that the roots of $B(x)$ are $-r_1, \ldots, -r_t \in \mathbb{Q} \setminus \mathbb{Z}$ and $-\alpha_1, \ldots, -\alpha_k \notin \mathbb{Q}$ with $k \geq 1$. Let N be the maximum order of all the irrational roots and suppose that for distinct α_i, α_j of order N, $\alpha_i \pm \alpha_j \notin \mathbb{Q}$. If the Gelfond–Schneider conjecture is true, then the series

$$\sum_{n \in \mathbb{Z}} \frac{A(n)}{B(n)}$$

and π are algebraically independent.

Proof. The case that $N = 1$ is dealt with in Theorem 24.1, so assume that $N > 1$. Let $v_1, \ldots, v_t$ and $m_1, \ldots, m_k$ be the orders of the roots respectively. By partial fractions we have

$$\frac{A(x)}{B(x)} = \sum_{j=1}^{k} \sum_{l=1}^{m_j} \frac{c_{j,l}}{(x+\alpha_j)^l} + \sum_{s=1}^{t} \sum_{u=1}^{v_s} \frac{d_{s,u}}{(x+r_s)^u}$$

for some algebraic numbers $c_{j,l}, d_{s,u}$. By Lemma 24.5 the series

$$\sum_{n \in \mathbb{Z}} A(n)/B(n)$$

equals

$$\pi i \sum_{j=1}^{k} c_{j,1} \frac{e^{2\pi i \alpha_j} + 1}{e^{2\pi i \alpha_j} - 1}$$

$$+ \sum_{j=1}^{k} \sum_{l=2}^{m_j} \frac{c_{j,l}(-1)^{l-1}(2\pi i)^l}{(l-1)!} \left(\frac{A_{l,1}}{e^{2\pi i \alpha_j} - 1} + \cdots + \frac{A_{l,l}}{(e^{2\pi i \alpha_j} - 1)^l} \right)$$

$$+ \pi i \sum_{s=1}^{t} d_{s,1} \frac{e^{2\pi i r_s} + 1}{e^{2\pi i r_s} - 1}$$

$$+ \sum_{s=1}^{t} \sum_{u=2}^{v_s} \frac{d_{s,u}(-1)^{u-1}(2\pi i)^u}{(u-1)!} \left(\frac{A_{u,1}}{e^{2\pi i r_s} - 1} + \cdots + \frac{A_{u,u}}{(e^{2\pi i r_s} - 1)^u} \right).$$

We view this sum as a polynomial in π. We examine the coefficient of π^N. Note that the rational roots contribute algebraic numbers to this coefficient so we ignore them for now. We focus on the transcendental portion of this coefficient which comes from the irrational roots part of the above sum. That is, ignoring the common factor of $\frac{(-1)^{N-1}(2i)^N}{(N-1)!}$, we examine

$$\sum_{\mathrm{ord}(\alpha_j)=N} c_{j,N} \left(\frac{A_{N,1}}{e^{2\pi i \alpha_j} - 1} + \cdots + \frac{A_{N,N}}{(e^{2\pi i \alpha_j} - 1)^N} \right).$$

We proceed similar to the proof of Theorem 24.1 and let

$$M, \beta, d, n_{a,j}, \alpha, \gamma_j, x_a, X_a, y$$

and $\bar{e}$ be as described there. By showing that there is an $\bar{e}$ so that the function

$$F(y) = \sum_{\mathrm{ord}(\alpha_j)=N} c_{j,N} \left(\frac{A_{N,1}}{\gamma_j y^{\overline{n_j} \cdot \bar{e}} - 1} + \cdots + \frac{A_{N,N}}{(\gamma_j y^{\overline{n_j} \cdot \bar{e}} - 1)^N} \right)$$

$$= \sum_{\mathrm{ord}(\alpha_j)=N} c_{j,N} \left(\frac{A_{N,1}(\gamma_j y^{\overline{n_j} \cdot \bar{e}} - 1)^{N-1} + \cdots + A_{N,N}}{(\gamma_j y^{\overline{n_j} \cdot \bar{e}} - 1)^N} \right)$$

is not constant, we show that the original coefficient of π^N is transcendental. By the remarks made above (24.2), we can assume that each $\overline{n_j} \cdot \overline{e}$ is positive (or else we could remove an algebraic number as we see in (24.2)). Note that we can choose $\overline{e}$ such that each $\overline{n_j} \cdot \overline{e}$ is distinct and non-zero. Thus, after placing everything over a common denominator, we have a function in y whose numerator has smaller degree than the denominator. If this function is constant (and therefore equal to zero), it is easy to see that this implies that each $c_{j,N}$ is zero which is a contradiction. Thus the coefficient of π^N is transcendental. Write

$$S = \sum_{n \in \mathbb{Z}} \frac{A(n)}{B(n)} = C_N \pi^N + \cdots + C_1 \pi$$

where each $C_i \in \overline{\mathbb{Q}}(\alpha^\beta, \dots, \alpha^{\beta^{d-1}})$ and $C_N \notin \overline{\mathbb{Q}}$. Similar to before, the Gelfond–Schneider conjecture implies algebraic independence of π and the coefficients, C_j, thus S is transcendental and in fact algebraically independent with π. $\square$

In the case of multiple roots, these methods allow one to obtain the following theorem. It can be viewed as a natural generalisation of Euler's classical theorem that $\zeta(2k) \in \pi^{2k}\mathbb{Q}$, where $\zeta(s)$ is the Riemann zeta function. As before, let $K_1 = \mathbb{Q}(\alpha_1, \dots, \alpha_k)$, the field generated by the roots of $B(x)$, and K_2 be K_1 adjoin the coefficients of $A(x)$ and $B(x)$.

Theorem 24.8 *Let $A(x), B(x)$ be polynomials with algebraic coefficients, $\deg A < \deg B$, and $A(x)$ is co prime to $B(x)$. Let K_1 be either an imaginary quadratic field or $\mathbb{Q}$. If $B(x)$ has no integral roots, then*

$$\sum_{n \in \mathbb{Z}}' \frac{A(n)}{B(n)}$$

is either zero or transcendental. If $B(x)$ has at least one integral root, then the sum is either in K_2 or transcendental. If $B(x)$ has at least one irrational root and all irrational roots satisfy the conditions of Theorem 24.1 that $\alpha_i \pm \alpha_j \notin \mathbb{Q}$ for $i \neq j$, then the sum is transcendental.

Proof. Suppose first that $K_1 = \mathbb{Q}$ and that $B(x)$ has no integral roots. Using (24.5), the sum of the series is $\pi P(\pi)$ for some polynomial $P(x) \in \overline{\mathbb{Q}}[x]$. If $P(x)$ is identically zero, the sum is zero. If $P(x)$ is not identically zero, then, the sum is a non-constant polynomial in π and hence transcendental. Suppose now that K_1 is an imaginary quadratic field $\mathbb{Q}(\sqrt{-D})$ with $D > 0$ and $B(x)$ has no integral roots. Again using (24.5) and the identity $\sum_{n \in \mathbb{Z}} \frac{1}{(n+x)} = \pi i \frac{e^{2\pi i x} + 1}{e^{2\pi i x} - 1}$, our sum is of the form

$$\pi R(\pi, e^{\pi \sqrt{D}/M})$$

where $R(x, y)$ is a rational function with algebraic coefficients which is polynomial in x and M is the same as was defined in the proof of Theorem 24.1. If $R(x, y)$ is identically zero, the sum is zero. If it is not identically zero, by

Nesterenko's theorem, it is transcendental since π and $e^{\pi\sqrt{D}}$ are algebraically independent. This completes the first part of the proof.

To treat the case that $B(x)$ may have integer roots, we argue as in the earlier theorems. In this context, we inject the observation made earlier with the three sums (24.6) from which it was deduced that the sum in question is of the form

$$\pi P(\pi) + \pi R(\pi, e^{\pi\sqrt{D}/M}) + \text{algebraic number},$$

where the algebraic number lies in the field K_2 being essentially a finite sum of terms of the form

$$\frac{c_{j,l}}{(n_t + \alpha_j)^l}, \quad \frac{d_{s,u}}{(n_t + r_s)^u}, \quad \frac{e_{p,q}}{(n_t - n_p)^q}$$

where n_t is an integral root, α_j is an irrational root, r_s is a rational root, n_p is an integral root not equal to n_t, and $c_{j,l}, d_{s,u}, e_{p,q}$ are the coefficients arising from the partial fractions decomposition of $A(x)/B(x)$. It is clear that the algebraic number is an element of K_2. Thus, if $P(x) + R(x,y) = 0$, then the sum is in K_2, otherwise the sum is transcendental by the earlier argument using Nesterenko's Theorem.

Finally, if the irrational roots of $B(x)$ satisfy the conditions of Theorem 24.1, then $R(x,y)$ depends on the variable y in which case we can conclude the sum is transcendental. $\square$

There are easily identifiable situations when one can say definitively that the sum is transcendental. For example, as in Theorem 24.8, if the irrational roots of $B(x)$ satisfy the conditions of Theorem 24.1 and generate an imaginary quadratic field, then the sum is transcendental. But there are other cases when the conditions of Theorem 24.1 may not be satisfied and still, one can check directly that the sum is transcendental. (See for example, Exercise 2 below.)

Another illustration is given by a problem investigated by Bundschuh. In 1979, Bundschuh [26] studied the series

$$\sum_{|n|\geq 2} \frac{1}{n^k - 1} \tag{24.8}$$

and showed using Schanuel's conjecture that all of these sums are transcendental numbers for $k \geq 3$. An examination of his proof shows that the "weaker" Gelfond–Schneider conjecture is sufficient to deduce his result. The methods described in this chapter allows one to deduce unconditionally that the sum (24.8) is transcendental for $k = 3, 4$ and 6.

Exercises

1. Show that

$$\sum_{n\in\mathbb{Z}} \frac{2n-1}{n^2 - n - 1} = 0.$$

2. Prove that

$$\sum_{n\in\mathbb{Z}}\frac{1}{An^2+Bn+C}=\frac{2\pi}{\sqrt{D}}\left(\frac{e^{2\pi\sqrt{D}/A}-1}{e^{2\pi\sqrt{D}/A}-2(\cos(\pi B/A))e^{\pi\sqrt{D}/A}+1}\right)$$

is transcendental if $A,B,C\in\mathbb{Z}$ and $-D=B^2-4AC<0$. Deduce that the value of the sum is a transcendental number.

3. From the previous exercise, deduce by taking appropriate limits that $\zeta(2)=\pi^2/6$.

4. Deduce a formula for

$$\sum_{n\in\mathbb{Z}}\frac{1}{(An^2+Bn+C)^k}$$

by treating the sum as a function of a continuous variable C and differentiating the right-hand side with respect to C.

5. Show that at least one of

$$\sum_{n=2}^{\infty}\frac{1}{n^3+1}\quad\text{or}\quad\sum_{n-2}^{\infty}\frac{1}{n^3-1},$$

is transcendental.

6. Show that for $k=2$, the sum (24.8) is a telescoping sum equal to $3/2$.

7. Show that for $k=4$, the sum (24.8) is equal to

$$\frac{7}{4}-\frac{\pi}{2}\coth\pi,$$

and that it is transcendental.

Chapter 25

Linear Independence of Values of Dirichlet L-Functions

We have seen before that for any non-trivial Dirichlet character $\chi \bmod q$, $L(1, \chi)$ is transcendental. In this chapter, we study the possible $\overline{\mathbb{Q}}$-linear relations between these values of $L(1, \chi)$ as χ ranges over all non-trivial Dirichlet characters $\bmod q$ with $q > 2$. More precisely, following [97], we will prove the following:

Theorem 25.1 *The $\overline{\mathbb{Q}}$-vector space generated by the values $L(1, \chi)$ as χ ranges over the non-trivial Dirichlet characters (mod q) has dimension $\varphi(q)/2$.*

We note that the analogous question for the dimension of the $\mathbb{Q}$-vector space generated by these special values is not yet resolved except in certain special cases. For instance, when $(q, \varphi(q)) = 1$, we know that the dimension of the $\mathbb{Q}$-vector space generated by these L-values is $\varphi(q) - 1$.

An important ingredient in the proof of the above theorem is the properties of a set of real multiplicatively independent units in the cyclotomic field discovered by K. Ramachandra (see Theorem 8.3 on p. 147 of [130] as well as [103]). We shall give the details along the course of the proof.

We begin with a straightforward result from group theory which is an interesting variant of Artin's theorem on the linear independence of the irreducible characters of a finite group G. As usual, we can define an inner product on the space $C(G)$ of complex-valued functions on G. Indeed, if $f, g \in C(G)$, then

$$(f, g) = \frac{1}{|G|} \sum_{x \in G} f(x)\overline{g(x)}.$$

M.R. Murty and P. Rath, *Transcendental Numbers*, DOI 10.1007/978-1-4939-0832-5_25, 153
© Springer Science+Business Media New York 2014

Lemma 25.2 *Let G be a finite group. Suppose that*

$$\sum_{\chi \neq 1} \chi(g)u_\chi = 0$$

for all $g \neq 1$ and all irreducible characters $\chi \neq 1$ of G. Then $u_\chi = 0$ for all $\chi \neq 1$.

Proof. For any irreducible character $\psi \neq 1$, we can multiply our equation by $\overline{\psi(g)}/|G|$ and sum over $g \neq 1$ to obtain

$$0 = \frac{1}{|G|} \sum_{g \neq 1} \overline{\psi(g)} \sum_{\chi \neq 1} \chi(g)u_\chi = \sum_{\chi \neq 1} u_\chi \left((\chi, \psi) - \frac{\psi(1)\chi(1)}{|G|} \right).$$

Thus, by the orthogonality relations,

$$0 = u_\psi - \frac{\psi(1)}{|G|} \sum_{\chi \neq 1} u_\chi \chi(1) = u_\psi - \frac{\psi(1)}{|G|} S \quad \text{(say)}.$$

Hence, for every $g \neq 1$, we have

$$0 = \frac{1}{|G|} \sum_{\chi \neq 1} \chi(g)\chi(1)S.$$

Recalling that

$$\frac{1}{|G|} \sum_{\chi} \chi(g)\chi(1) = 0$$

unless $g = 1$, we deduce that $S = 0$. Hence $u_\chi = 0$ for all $\chi \neq 1$ as desired. $\square$

Let us record the following version of Baker's theorem which is amenable for our applications.

Lemma 25.3 *If $\alpha_1, \ldots, \alpha_n \in \overline{\mathbb{Q}}\backslash\{0\}$ and $\beta_1, \ldots, \beta_n \in \overline{\mathbb{Q}}$, then*

$$\beta_1 \log \alpha_1 + \cdots + \beta_n \log \alpha_n$$

is either zero or transcendental. The latter case arises if $\log \alpha_1, \ldots, \log \alpha_n$ are linearly independent over $\mathbb{Q}$ and $\beta_1, \ldots, \beta_n$ are not all zero.

As before, for all purposes we interpret log as the principal value of the logarithm with the argument lying in the interval $(-\pi, \pi]$.

In particular, if $\log \alpha_1, \ldots, \log \alpha_n$ are linearly independent over $\mathbb{Q}$, then they are linearly independent over $\overline{\mathbb{Q}}$.

As an application of the above lemma, we first prove the following:

Lemma 25.4 *Let $\alpha_1, \alpha_2, \ldots, \alpha_n$ be positive algebraic numbers. If $c_0, c_1, \ldots, c_n$ are algebraic numbers with $c_0 \neq 0$, then*

$$c_0 \pi + \sum_{j=1}^{n} c_j \log \alpha_j$$

is a transcendental number and hence non-zero.

Proof. Let S be such that $\{\log \alpha_j : j \in S\}$ is a maximal $\mathbb{Q}$-linearly independent subset of

$$\log \alpha_1, \ldots, \log \alpha_n.$$

We write $\pi = -i \log(-1)$. We can re-write our linear form as

$$-ic_0 \log(-1) + \sum_{j \in S} d_j \log \alpha_j,$$

for algebraic numbers d_j. By Baker's theorem, this is either zero or transcendental. The former case cannot arise if we show that

$$\log(-1), \quad \log \alpha_j, \quad j \in S$$

are linearly independent over $\mathbb{Q}$. But this is indeed the case since

$$b_0 \log(-1) + \sum_{j \in S} b_j \log \alpha_j = 0$$

for integers $b_0, b_j, j \in S$ implies that b_0 is necessarily zero. This is because the sum $\sum_{j \in S} b_j \log \alpha_j$ is a real number since each α_j is a positive real number. But then $b_j = 0$ for all j. This completes the proof. $\square$

One of the crucial ingredients for the proof of our theorem is the following:

Theorem 25.5 *Let f_e be an even algebraic valued periodic function defined over integers with period q. Suppose it is supported at co-prime classes (mod q) with $\sum_{a=1}^{q} f_e(a) = 0$. Then $L(1, f_e) \neq 0$ unless f_e is identically zero. Moreover, $L(1, f_e)$ is an algebraic linear combination of logarithms of multiplicatively independent units of the q-th cyclotomic field. In particular, if f_e is not identically zero, then $L(1, f_e)$ is transcendental.*

Proof. The proof involves a different approach to the original problem of Chowla. The strategy is to write any periodic function as the sum of an even and odd function. Let us consider an algebraic-valued function f supported on the co-prime classes (mod q) with

$$\sum_{a=1}^{q} f(a) = 0.$$

We want to write f as

$$f = f_e + f_o$$

where f_e is even (i.e. $f_e(-n) = f_e(n)$) and f_o is odd (i.e. $f_o(-n) = -f_o(n)$). Since the characters form a basis for the space of such functions, we can write

$$f = \sum_{\chi \neq 1} c_\chi \chi.$$

Here the sum is over non-trivial Dirichlet characters χ (mod q). Note that the trivial character is absent because

$$\sum_{a=1}^{q} f(a) = 0.$$

Thus, we obtain the desired decomposition for f by considering

$$f_e = \sum_{\chi \text{ even}, \chi \neq 1} c_\chi \chi,$$

and

$$f_o = \sum_{\chi \text{ odd}} c_\chi \chi.$$

We recall that Ramachandra (see Theorem 8.3 on p. 147 of [130] as well as [103]) discovered a set of real multiplicatively independent units in the cyclotomic field. Let us denote these units by ξ_a (with $1 < a < q/2$ and $(a, q) = 1$) following the notation of [130]. A fundamental property of these units is that it enables us to obtain an expression for $L(1, \chi)$ for an even non-trivial character χ which is amenable for applying Baker's theory. More precisely, one has the following formula: for even χ with $\chi \neq 1$, we have

$$L(1, \chi) = A_\chi \sum_{1 < a < q/2} \overline{\chi}(a) \log \xi_a, \qquad (25.1)$$

where A_χ is a non-zero algebraic number. See the proof of Theorem 8.3 on p. 149 in [130] for deriving this expression. We note that this can be regarded as the cyclotomic analogue of one of the main theorems of [102] in the case of an imaginary quadratic field (see also [103]).

To elaborate, let ζ be a primitive q-th root of unity and following Ramachandra [102], define

$$\eta_a = \prod_{d \mid q, d \neq q, (d, q/d) = 1} \frac{1 - \zeta^{ad}}{1 - \zeta^d}.$$

Setting

$$d_a = \frac{1}{2}(1 - a) \sum_{d \mid q, (d, q/d) = 1, d \neq q} d,$$

one sees that $\xi_a = \zeta^{d_a} \eta_a$ lies in the real subfield $\mathbb{Q}(\zeta + \zeta^{-1})$. These are the multiplicatively independent units for $1 < a < q/2$ with $(a, q) = 1$. Following the calculation on p. 149 in [130], we see that

$$\sum_{a=1}^{q} \overline{\chi}(a) \sum_{d \mid q, (d, q/d) = 1, d \neq q} \log |1 - \zeta_q^{ad}|$$

is a non-zero algebraic multiple of $L(1,\chi)$. This easily leads to the formula (25.1) above. For a more detailed theory, we direct the reader to Theorems 8, 9 and 12 in [102] (see also [98] for an application in a different set-up). Thus we have

$$L(1, f_e) = \sum_{\chi \text{ even}, \chi \neq 1} c_\chi L(1,\chi)$$

$$= \sum_{\chi \text{ even}, \chi \neq 1} c_\chi A_\chi \left(\sum_{1 < a < q/2} \overline{\chi}(a) \log \xi_a \right)$$

$$= \sum_{1 < a < q/2} \left(\sum_{\chi \text{ even}, \chi \neq 1} A_\chi c_\chi \overline{\chi}(a) \right) \log \xi_a.$$

Since the ξ_a's are multiplicatively independent, the $\log \xi_a$'s are linearly independent over $\mathbb{Q}$. By Baker's theorem, they are linearly independent over $\overline{\mathbb{Q}}$. Consequently, $L(1, f_e) = 0$ if and only if

$$\sum_{\chi \text{ even}, \chi \neq 1} A_\chi c_\chi \overline{\chi}(a) = 0, \qquad 1 < a < q/2.$$

Now the even characters of $(\mathbb{Z}/q\mathbb{Z})^*$ can be viewed as characters of the group $(\mathbb{Z}/q\mathbb{Z})^*/\{\pm 1\}$. Thus by Lemma 25.2 and since $A_\chi \neq 0$, we deduce that $c_\chi = 0$ for all even χ. This proves the theorem. $\square$

As an immediate corollary, we deduce the following result in the classical case:

Corollary 25.6 $L(1,\chi)$, as χ ranges over non-trivial even characters mod q, are linearly independent over $\overline{\mathbb{Q}}$.

We remark that the above corollary together with Schanuel's conjecture implies the algebraic independence of the $L(1,\chi)$ as χ ranges over the even Dirichlet characters mod q.

Finally, we shall need the following observation which we leave as an exercise.

Lemma 25.7 For any odd Dirichlet character χ, $L(1,\chi)$ is an algebraic multiple of π.

We can now prove the main result of the chapter. As noted above, for odd characters, each $L(1,\chi)$ is equal to an algebraic multiple of π. However for an even non-trivial character, as we have seen before, $L(1,\chi)$ is a non-zero algebraic multiple of

$$\sum_{a=1}^{q} \overline{\chi}(a) \sum_{d|q, (d,q/d)=1, d \neq q} \log |1 - \zeta_q^{ad}|.$$

Thus in view of Lemma 25.4, the $\overline{\mathbb{Q}}$-space generated by the even $L(1,\chi)$ values is linearly disjoint from that generated by the odd $L(1,\chi)$ values. Since for any odd character χ, $L(1,\chi) \neq 0$, this proves our main theorem.

Exercises

1. Prove that for any odd Dirichlet character χ mod q, $L(1,\chi)$ is an algebraic multiple of π.

2. Prove Artin's theorem that the irreducible characters of a finite group are linearly independent over the field of complex numbers.

3. Show that Schanuel's conjecture implies that the numbers $L(1,\chi)$, where χ ranges over the non-trivial even Dirichlet characters mod q, are algebraically independent.

4. Without appealing to the Ramachandra units, show directly that for any non-trivial even Dirichlet character χ mod q, $L(1,\chi)$ an algebraic linear combination of logarithms of positive algebraic numbers.

5. Show that for any prime p, the numbers $L(1,\chi)$, where χ runs over the non-trivial Dirichlet characters mod p, are linearly independent over $\mathbb{Q}$.

Chapter 26

Transcendence of Values of Class Group L-Functions

In this chapter, we consider the analog of the question discussed in Chap. 25 for class group L-functions. We refer to the original work [96] for further details.

Let K be an algebraic number field and f a complex-valued function of the ideal class group $\mathcal{H}_K$ of K. Here, we consider the Dirichlet series

$$L(s, f) := \sum_{\mathfrak{a}} \frac{f(\mathfrak{a})}{\mathrm{N}(\mathfrak{a})^s}, \tag{26.1}$$

where the summation is over all integral ideals $\mathfrak{a}$ of the ring of integers $\mathcal{O}_K$ of K. If f is identically 1, then $L(s, f)$ is the Dedekind zeta function of K. If f is a character χ of the ideal class group $\mathcal{H}_K$ of K, then $L(s, \chi)$ is a Hecke L-function.

Let us begin with the following theorem. We have seen a similar result in Chap. 22.

Theorem 26.1 $L(s, f)$ *extends analytically for all* $s \in \mathbb{C}$ *except possibly at* $s = 1$ *where it may have a simple pole with residue a non-zero multiple of*

$$\rho_f := \sum_{\mathfrak{a} \in \mathcal{H}_K} f(\mathfrak{a}).$$

The series (26.1) converges at $s = 1$ *if and only if* $\rho_f = 0$.

Proof. Since f is a function on the ideal class group, we have

$$L(s, f) = \sum_{\mathfrak{C} \in \mathcal{H}_K} f(\mathfrak{C}) \zeta(s, \mathfrak{C})$$

M.R. Murty and P. Rath, *Transcendental Numbers*, DOI 10.1007/978-1-4939-0832-5_26, 159
© Springer Science+Business Media New York 2014

where

$$\zeta(s, \mathfrak{C}) = \sum_{\mathfrak{a} \in \mathfrak{C}} \frac{1}{N(\mathfrak{a})^s}.$$

It is classical (see [76]) that each $\zeta(s, \mathfrak{C})$ extends to all $s \in \mathbb{C}$ with the exception of $s = 1$, where it has a simple pole with residue

$$\frac{2^{r_1}(2\pi)^{r_2} R_K}{w\sqrt{|d_K|}},$$

where r_1 is the number of real embeddings, $2r_2$ is the number of complex embeddings and R_K is the regulator of K. We conclude that $L(s, f)$ extends analytically to all $s \in \mathbb{C}$ apart from a simple pole at $s = 1$ with residue

$$\frac{2^{r_1}(2\pi)^{r_2} R_K}{w\sqrt{|d_K|}} \sum_{\mathfrak{C}} f(\mathfrak{C}).$$

Thus, $L(s, f)$ is analytic at $s = 1$ if and only if $\rho_f = 0$. To study the convergence of the Dirichlet series $L(s, f)$ at $s = 1$, we proceed as follows. The number of ideals with norm $\leq x$ and lying in a fixed class $\mathfrak{C}$ is well known to be (see [76]),

$$\frac{2^{r_1}(2\pi)^{r_2} R_K}{w\sqrt{|d_K|}} x + O(x^{\frac{d}{d+1}}), \qquad (26.2)$$

where d is the degree of K over $\mathbb{Q}$. Letting

$$S(x) = \sum_{N(\mathfrak{a}) \leq x} f(\mathfrak{a}),$$

we have by the general technique of partial summation (see p. 17 of [94]) that

$$L(s, f) = s \int_1^\infty \frac{S(x)}{x^{s+1}} dx,$$

for $\Re(s) > 1$. Now,

$$S(x) = \sum_{\mathfrak{C} \in \mathcal{H}_K} \sum_{\mathfrak{a} \in \mathfrak{C}, N(\mathfrak{a}) \leq x} f(\mathfrak{a}) = \sum_{\mathfrak{C} \in \mathcal{H}_K} f(\mathfrak{C}) \left(\frac{2^{r_1}(2\pi)^{r_2} R_K}{w\sqrt{|d_K|}} x + O(x^{\frac{d}{d+1}}) \right)$$

which is easily seen to be

$$\frac{2^{r_1}(2\pi)^{r_2} R_K \rho_f}{w\sqrt{|d_K|}} x + O(x^{\frac{d}{d+1}}).$$

Hence, by partial summation, it follows immediately that the Dirichlet series $L(s, f)$ converges at $s = 1$ if and only if $\rho_f = 0$. This completes the proof. $\square$

Thus in the case that the series converges at $s = 1$, it makes sense to consider the Dirichlet series

$$L(1, f) = \sum_{\mathfrak{a}} \frac{f(\mathfrak{a})}{\mathbf{N}(\mathfrak{a})}.$$

Our goal in this chapter is to investigate special values of $L(s, f)$ at $s = 1$ when K is an imaginary quadratic field and f takes algebraic values. In particular, we will investigate the transcendental nature of $L(1, \chi)$ when χ is an ideal class character.

The basic tools are Kronecker's limit formula and Baker's theory of linear forms in logarithms. In particular, we will show that the values $L(1, \chi)$ are linearly independent over $\overline{\mathbb{Q}}$ as χ ranges over non-trivial characters of the ideal class group (modulo the action of complex conjugation on the group of characters). This is analogous to the result we discussed in Chap. 25.

We will use Kronecker's limit formula as discussed in the works of Siegel [113], Ramachandra [102] and Lang [77]. We cannot give an in-depth discussion about these topics, but shall be content with a brief review.

Let $\Delta(z)$ be the discriminant function:

$$\Delta(z) = (2\pi)^{12} \, q \prod_{n=1}^{\infty} (1 - q^n)^{24} = (2\pi)^{12} \eta(z)^{24}, \qquad q = e^{2\pi i z}.$$

Here η^{24} is the Ramanujan cusp form.

Now let K be an imaginary quadratic field and let $\mathfrak{b}$ be an ideal of $\mathcal{O}_K$. If $[\beta_1, \beta_2]$ is an integral basis of $\mathfrak{b}$ with $\Im(\beta_2 / \beta_1) > 0$, we define

$$g(\mathfrak{b}) = (2\pi)^{-12} (\mathbf{N}(\mathfrak{b}))^6 \Delta(\beta_1, \beta_2),$$

where

$$\Delta(\omega_1, \omega_2) := \omega_1^{-12} \Delta\left(\frac{\omega_2}{\omega_1}\right),$$

One can verify (as on p. 109 of [102]) that $g(\mathfrak{b})$ is well defined and does not depend on the choice of integral basis of $\mathfrak{b}$. Furthermore, $g(\mathfrak{b})$ depends only on the ideal class $\mathfrak{b}$ belongs to in the ideal class group (see Lemma 2 of [102], also p. 280 of [77]). Thus, if $\mathfrak{C}$ is an ideal class, we write $g(\mathfrak{C})$ for the common value $g(\mathfrak{b})$ as $\mathfrak{b}$ ranges over the elements of the class $\mathfrak{C}$.

Let d_K be the discriminant of K and w denote the number of roots of unity in $\mathcal{O}_K$. As before, writing

$$\zeta(s, \mathfrak{C}) = \sum_{\mathfrak{a} \in \mathfrak{C}} \frac{1}{\mathbf{N}(\mathfrak{a})^s}$$

for the ideal class zeta function, we have by Kronecker's limit formula

$$\zeta(s, \mathfrak{C}) = \frac{2\pi}{w\sqrt{|d_K|}} \left(\frac{1}{s-1} + 2\gamma - \log|d_K| - \frac{1}{6}\log|g(\mathfrak{C}^{-1})| \right) + O(s-1), \quad (26.3)$$

as $s \to 1^+$. (Note that there is a sign error in formula (2) on p. 280 of [77].)

Proposition 26.2 *If $\mathfrak{C}_1$ and $\mathfrak{C}_2$ are ideal classes, then $g(\mathfrak{C}_1)/g(\mathfrak{C}_2)$ is an algebraic number lying in the Hilbert class field of K.*

Proof. This follows immediately from Lemma 3 of [30] and is a classical result from the theory of complex multiplication. $\square$

Furthermore, we have

Proposition 26.3 *Let K_H be the Hilbert class field of K. Now if $\mathfrak{p}$ is a prime ideal of K and $\sigma_{\mathfrak{p}}$ is the Frobenius automorphism in $\mathrm{Gal}(K_H/K)$, then for any ideal $\mathfrak{b}$, $g(\mathfrak{b})/g(\mathcal{O}_K) \in K_H$ and we have*

$$\sigma_{\mathfrak{p}}\left(g(\mathfrak{b})/g(\mathcal{O}_K)\right) = g(\mathfrak{p}^{-1}\mathfrak{b})/g(\mathfrak{p}^{-1}\mathcal{O}_K), \quad \overline{g(\mathfrak{b})/g(\mathcal{O}_K)} = g(\mathfrak{b}^{-1})/g(\mathcal{O}_K).$$

Proof. The first part is the content of Theorem 1 on p. 161 of [77]. The action of complex conjugation is easily deduced from the equation $\bar{j}(\mathfrak{b}) = j(\bar{\mathfrak{b}})$ for the j-function $\square$

For imaginary quadratic fields, by a deeper analysis we will show the following:

Theorem 26.4 *Let K be an imaginary quadratic field and $f : \mathcal{H}_K \to \overline{\mathbb{Q}}$ be not identically zero. Suppose that $\rho_f = 0$. Then, $L(1,f) \neq 0$ unless $f(\mathfrak{C}) + f(\mathfrak{C}^{-1}) = 0$ for every ideal class $\mathfrak{C} \in \mathcal{H}_K$. Moreover, $L(1,f)/\pi$ is a $\overline{\mathbb{Q}}$-linear combination of logarithms of algebraic numbers. In particular, $L(1,f)/\pi$ is transcendental whenever $L(1,f) \neq 0$.*

This result has several interesting corollaries. Before giving a proof of the above theorem, let us first derive these consequences.

Corollary 26.5 *Let K be an imaginary quadratic field and χ a non-trivial character of the ideal class group of K. Then, $L(1,\chi)/\pi$ is a non-zero $\overline{\mathbb{Q}}$-linear combination of logarithms of algebraic numbers and hence transcendental.*

Proof. To prove this corollary, we begin by noting that in the case K is an imaginary quadratic field, the formulas become simple and we can apply Kronecker's limit formula. In this situation, when the series converges, we have by (26.3)

$$\frac{L(1,f)}{\pi} = \frac{-1}{3w\sqrt{|d_K|}} \sum_{\mathfrak{C} \in \mathcal{H}_K} f(\mathfrak{C}) \log |g(\mathfrak{C}^{-1})|. \tag{26.4}$$

Now we invoke Proposition 26.2. Indeed, by this proposition, we have for the identity class $\mathfrak{C}_0$, that $g(\mathfrak{C}^{-1})/g(\mathfrak{C}_0)$ is algebraic. Thus, as $\rho_f = 0$, we have

$$\frac{L(1,f)}{\pi} = \frac{-1}{3w\sqrt{|d_K|}} \sum_{\mathfrak{C} \in \mathcal{H}_K} f(\mathfrak{C}) \log |g(\mathfrak{C}^{-1})/g(\mathfrak{C}_0)|, \tag{26.5}$$

for any fixed class $\mathfrak{C}_0$ of $\mathcal{H}_K$. Specialising to the case $f = \chi$, where χ is a non-trivial character of the ideal class group $\mathcal{H}_K$, and using the theorem that

$L(1,\chi) \neq 0$, we deduce Corollary 26.5 by virtue of Baker's theorem. This completes the proof. $\square$

Since complex conjugation acts on the group of ideal class characters we see by a simple calculation that $L(1,\chi) = L(1,\overline{\chi})$ for any ideal class character χ. We denote by $\mathcal{H}_K^*$ a set of orbit representatives under this action. Now we have:

Corollary 26.6 *Let K be an imaginary quadratic field and $\mathcal{H}_K$ its ideal class group. The values $L(1,\chi)$ (as χ ranges over the non-trivial characters of $\mathcal{H}_K^*$) and π are linearly independent over $\overline{\mathbb{Q}}$.*

Proof. Suppose that

$$\sum_{\chi\neq 1, \chi\in\mathcal{H}_K^*} c_\chi L(1,\chi) \in \overline{\mathbb{Q}}\pi,$$

for some $c_\chi \in \overline{\mathbb{Q}}$. Then, setting

$$f = \sum_{\chi\neq 1, \chi\in\mathcal{H}_K^*} c_\chi \chi,$$

we have $L(1,f)/\pi$ is algebraic. Since $\rho_f = 0$, we can apply Theorem 26.4 and deduce that f is identically zero. By the independence of characters, this means that each c_χ is zero. $\square$

Thus in the special case that the ideal class group $\mathcal{H}_K$ is an elementary abelian 2-group, the corollary implies that the $L(1,\chi)$ as χ ranges over the non-trivial characters of $\mathcal{H}_K$ are linearly independent over $\overline{\mathbb{Q}}$.

Theorem 26.4 also implies that at most one such $L(1,\chi)$ is algebraic. Indeed, the following corollary follows directly from Corollary 26.6 since two algebraic numbers are linearly dependent over $\overline{\mathbb{Q}}$.

Corollary 26.7 *All of the values $L(1,\chi)$ as χ ranges over the non-trivial characters of $\mathcal{H}_K^*$, with at most one exception, are transcendental.*

The elimination of this singular possibility, in other words, the proof of transcendence of $L(1,\chi)$ for all non-trivial χ seems difficult and is related to Schanuel's conjecture. Indeed, a "weaker" version of Schanuel suffices for our purposes. This is the conjecture that logarithms of algebraic numbers which are linearly independent over $\mathbb{Q}$ are algebraically independent. Assuming the "weaker" Schanuel's conjecture which was discussed in an earlier chapter, one can show the transcendence of $L(1,\chi)$ for all non-trivial χ.

We now come to the proof of Theorem 26.4. In view of (26.5) and Baker's theorem, the only part of Theorem 26.4 that remains to be proved is the non-vanishing of $L(1,f)$ subject to the hypothesis of the given theorem. To this end, we will require three lemmas.

Lemma 26.8 *Let K be an imaginary quadratic field and $f : \mathcal{H}_K \to \overline{\mathbb{Q}}$. Then, $L(1,f) = 0$ implies that $L(1,f^\sigma) = 0$ for any Galois automorphism σ of $\mathrm{Gal}(\overline{\mathbb{Q}}/\mathbb{Q})$.*

Proof. Equation (26.5) expresses $L(1, f)/\pi$ as a linear form of logarithms of algebraic numbers. Now choose a maximal set of $\mathbb{Q}$-linearly independent numbers from $\{\log |g(\mathfrak{C}^{-1})/g(\mathfrak{C}_0)| : \mathfrak{C} \in \mathcal{H}_K\}$. Denote this set by $\log \alpha_1, \ldots, \log \alpha_t$. Thus,

$$\log |g(\mathfrak{C}^{-1})/g(\mathfrak{C}_0)| = \sum_{j=1}^{t} x(\mathfrak{C}, j) \log \alpha_j,$$

where the $x(\mathfrak{C}, j)$'s are rational numbers. Hence

$$L(1, f) = -\frac{\pi}{3w\sqrt{|d_K|}} \sum_{j=1}^{t} \sum_{\mathfrak{C} \in \mathcal{H}_K} f(\mathfrak{C}) x(\mathfrak{C}, j) \log \alpha_j.$$

If $L(1, f) = 0$, then Baker's theorem gives that

$$\sum_{\mathfrak{C} \in \mathcal{H}_K} f(\mathfrak{C}) x(\mathfrak{C}, j) = 0, \quad j = 1, 2, \ldots, t.$$

Since the $x(\mathfrak{C}, j)$'s are rational numbers, we deduce that for every Galois automorphism σ,

$$\sum_{\mathfrak{C} \in \mathcal{H}_K} f^{\sigma}(\mathfrak{C}) x(\mathfrak{C}, j) = 0, \quad j = 1, 2, \ldots, t.$$

Consequently, $L(1, f^{\sigma}) = 0$. $\square$

The next lemma allows us to reduce the proof of Theorem 26.4 to the case when f is rational-valued.

Lemma 26.9 *Let M be the algebraic number field generated by the values of f. Let $r = [M : \mathbb{Q}]$ and choose a basis $\beta_1, \ldots, \beta_r$ of M over $\mathbb{Q}$ and write*

$$f(\mathfrak{C}) = \sum_{i=1}^{r} \beta_i f_i(\mathfrak{C}),$$

with $f_i(\mathfrak{C})$ rational. Then, $L(1, f) = 0$ implies $L(1, f_i) = 0$ for $i = 1, 2, \ldots, r$.

Proof. Let $M = M^{(1)}, \ldots, M^{(r)}$ be the conjugate fields of M. The map $x \to x^{(j)}$ from M to $M^{(j)}$ extends to a Galois automorphism σ_j of $\mathrm{Gal}(\overline{\mathbb{Q}}/\mathbb{Q})$. Thus,

$$f^{\sigma_j}(\mathfrak{C}) = \sum_{i=1}^{r} \beta_i^{(j)} f_i(\mathfrak{C}).$$

Clearly, the matrix $(\beta_i^{(j)})$ is invertible since $\beta_1, \ldots, \beta_r$ is a basis, and we can express $f_i(\mathfrak{C})$ as a $\overline{\mathbb{Q}}$-linear combination of the $f^{\sigma_j}(\mathfrak{C})$. By Lemma 26.8, we have that $L(1, f) = 0$ implies $L(1, f^{\sigma_j}) = 0$ for every j. Thus, $L(1, f_i) = 0$ for $1 \leq i \leq r$, as desired. $\square$

Lemma 26.10 *If f is rational-valued and $L(1, f) = 0$, then $f(\mathfrak{C}) + f(\mathfrak{C}^{-1}) = 0$ for every ideal class $\mathfrak{C}$.*

Proof. If $L(1, f) = 0$, then

$$\sum_{\mathfrak{C} \in \mathcal{H}_K} f(\mathfrak{C}) \log |g(\mathfrak{C}^{-1})/g(\mathfrak{C}_0)| = 0.$$

Clearing denominators, we may suppose that f is integer-valued. Exponentiating the above expression gives

$$\prod_{\mathfrak{C} \in \mathcal{H}_K} \left| \frac{g(\mathfrak{C}^{-1})}{g(\mathfrak{C}_0)} \right|^{f(\mathfrak{C})} = 1.$$

To remove the absolute values, we square the expression and pair up $\mathfrak{C}$ with $\mathfrak{C}^{-1}$ and re-arrange it to deduce that

$$\prod_{\mathfrak{C}} \left[\frac{g(\mathfrak{C})}{g(\mathfrak{C}_0)} \right]^{f(\mathfrak{C}) + f(\mathfrak{C}^{-1})} = 1.$$

Each of the factors in the product is an algebraic number and applying Proposition 26.3, we see that

$$\prod_{\mathfrak{C}} \left[\frac{g(\mathfrak{p}^{-1} \mathfrak{C}^{-1})}{g(\mathfrak{p}^{-1} \mathfrak{C}_0)} \right]^{f(\mathfrak{C}) + f(\mathfrak{C}^{-1})} = 1,$$

for any prime ideal $\mathfrak{p}$ of $\mathcal{O}_K$. Taking absolute values and then logarithms, we conclude that

$$\sum_{\mathfrak{C}} (f(\mathfrak{C}) + f(\mathfrak{C}^{-1})) \log |g(\mathfrak{p}^{-1} \mathfrak{C}^{-1})/g(\mathfrak{p}^{-1} \mathfrak{C}_0)| = 0,$$

for every prime ideal $\mathfrak{p}$. By the Chebotarev density theorem, the $\mathfrak{p}^{-1}$'s range over all elements of $\mathcal{H}_K$ as $\mathfrak{p}$ ranges over all prime ideals of $\mathcal{O}_K$.

We view these equations as a matrix equation

$$DF = 0$$

where F is the transpose of the row vector $(f(\mathfrak{C}) + f(\mathfrak{C}^{-1}))_{\mathfrak{C} \in H_K}$ and D is the "Dedekind-Frobenius" matrix whose (i, j)-th entry is given by

$$\log g(\mathfrak{C}_i^{-1} \mathfrak{C}_j)/g(\mathfrak{C}_i^{-1})$$

with $\mathfrak{C}_i, \mathfrak{C}_j$ running over the elements of the ideal class group. The first column of D is the zero vector and we can re-write our matrix equation as

$$D_0 F_0 = 0$$

where F_0 is the transpose of the row vector $(f(\mathfrak{C}) + f(\mathfrak{C}^{-1}))_{\mathfrak{C} \neq 1}$ and D_0 is the matrix obtained from D by deleting the row and column corresponding to the identity element. By the theory of the Dedekind determinants discussed earlier (see also p. 71 of [130]), the determinant of D_0 is

$$\prod_{\chi \neq 1} \left(\sum_{\mathfrak{a}} \chi(\mathfrak{a}) \log g(\mathfrak{a}^{-1}) \right) \neq 0,$$

since by formula (26.4), each factor is up to a non-zero scalar, $L(1, \chi)$, which is non-zero. Thus, $f(\mathfrak{C}) + f(\mathfrak{C}^{-1}) = 0$ for all $\mathfrak{C} \neq \mathfrak{C}_0$. Since

$$\sum_{\mathfrak{C}} f(\mathfrak{C}) + f(\mathfrak{C}^{-1}) = 0,$$

we deduce that $f(\mathfrak{C}_0) + f(\mathfrak{C}_0^{-1}) = 2f(\mathfrak{C}_0) = 0$ as well. This completes the proof. $\square$

The proof of Theorem 26.4 can now be given as follows. First, if f is rational-valued, we are done by the previous lemma. Lemma 26.9 allows us to reduce to the rational-valued case. This completes the proof.

When χ is a genus character, one can relate $L(1, \chi)$ to classical Dirichlet L-functions attached to quadratic characters [113]. Let us first recall the relevant notions. As before, let K be an imaginary quadratic field with discriminant $D < 0$. Real-valued characters of the ideal class group of K are called genus characters. These characters can be extended to functions on the ideal classes of $\mathcal{O}_K$ in the obvious way. Such extended characters take on only the values $0, \pm 1$. By a classical theorem of Kronecker, they have a simple description. For each factorisation $D = D_1 D_2$ with D_1, D_2 being fundamental discriminants, we define a character χ_{D_1, D_2} by setting it to be

$$\chi_{D_1, D_2}(\mathfrak{p}) = \begin{cases} \chi_{D_1}(\mathbf{N}(\mathfrak{p})) & \text{if } (\mathfrak{p}, D_1) = 1 \\ \chi_{D_2}(\mathbf{N}(\mathfrak{p})) & \text{if } (\mathfrak{p}, D_2) = 1. \end{cases} \tag{26.6}$$

One can show that this is well defined and that it defines a character on the ideal class group. We refer the reader to p. 60 of [113] for the background on genus characters. We have the Kronecker factorisation formula:

$$L(s, \chi_{D_1, D_2}) = L(s, \chi_{D_1})L(s, \chi_{D_2}).$$

Corresponding to the factorisation $D = 1 \cdot D$, we get

$$L(s, \chi_{1, D}) = \zeta(s)L(s, \chi_D).$$

The left-hand side is $\zeta_K(s)$ and so we can write

$$\sum_{\mathfrak{C} \in \mathcal{H}_K} \zeta(s, \mathfrak{C}) = \zeta(s)L(s, \chi_D).$$

This identity could have been derived in other ways. Applying the Kronecker limit formula (26.3), and comparing the constant term in the Laurent expansion of both sides, we obtain as in [113]:

Proposition 26.11

$$\gamma L(1,\chi_D) + L'(1,\chi_D) = \frac{2\pi}{w\sqrt{|d_K|}} \sum_{\mathfrak{C}\in\mathcal{H}_K} \left(2\gamma - \log|d_K| - \frac{1}{6}\log|g(\mathfrak{C}^{-1})|\right).$$

Using Dirichlet's class number formula, we deduce:

Corollary 26.12

$$\frac{L'(1,\chi_D)}{L(1,\chi_D)} = \gamma - \log|d_K| - \frac{1}{6h} \sum_{\mathfrak{C}\in\mathcal{H}_K} \log|g(\mathfrak{C})|,$$

where h denotes the order of $\mathcal{H}_K$.

In particular, we deduce the following interesting result.

Theorem 26.13 *For any ideal class $\mathfrak{C}$,*

$$\frac{L'(1,\chi_D)}{L(1,\chi_D)} - \gamma + \frac{1}{6}\log|g(\mathfrak{C})|$$

is a $\overline{\mathbb{Q}}$-linear combination of logarithms of algebraic numbers.

We will make fundamental use of the following result of Nesterenko [86] which we recall again.

Proposition 26.14 *For any imaginary quadratic field with discriminant $-D$ and character χ_D, the numbers $\pi, e^{\pi\sqrt{D}}$ and*

$$\prod_{a=1}^{D} \Gamma(a/D)^{\chi_D(a)},$$

are algebraically independent over $\overline{\mathbb{Q}}$.

Now we have all the ingredients ready to prove the following.

Theorem 26.15 *Let K be an imaginary quadratic field with character χ_D. Then,*

$$\exp\left(\frac{L'(1,\chi_D)}{L(1,\chi_D)} - \gamma\right)$$

and π are algebraically independent. Here γ is Euler's constant.

Proof. We shall first analyse the asymptotic behaviour of the formula in Corollary 26.12 using the theory of Hurwitz zeta functions. Recall that the Hurwitz zeta function $\zeta(s,x)$ is defined by the series

$$\zeta(s,x) := \sum_{n=0}^{\infty} \frac{1}{(n+x)^s}.$$

This series converges for $\Re(s) > 1$ and Hurwitz showed how one can extend it to the entire complex plane apart from $s = 1$ where it has a simple pole with residue 1. Given a Dirichlet character χ mod q, we can write

$$L(s,\chi) = \sum_{n=1}^{\infty} \frac{\chi(n)}{n^s} = q^{-s} \sum_{a=1}^{q} \chi(a)\zeta(s, a/q).$$

Thus,

$$L'(s,\chi) = -(\log q)q^{-s} \sum_{a=1}^{q} \chi(a)\zeta(s, a/q) + q^{-s} \sum_{a=1}^{q} \chi(a)\zeta'(s, a/q).$$

Using the well-known formulas

$$\zeta(0,x) = \frac{1}{2} - x, \quad \zeta'(0,x) = \log(\Gamma(x)/2\pi),$$

where the differentiation is with respect to the s-variable, we deduce that

$$L(0,\chi) = \sum_{a=1}^{q} \chi(a)(1/2 - a/q),$$

and

$$L'(0,\chi) = -(\log q)L(0,\chi) + \sum_{a=1}^{q} \chi(a) \log \Gamma(a/q), \qquad (26.7)$$

since $\sum_{a=1}^{q} \chi(a) = 0$. If χ is odd and primitive, $L(s,\chi)$ satisfies a functional equation of the form

$$A^s \Gamma((s+1)/2)L(s,\chi) = \varpi A^{1-s} \Gamma((2-s)/2)L(1-s,\overline{\chi}),$$

where ϖ (called the root number) is a complex number (see p. 71 of [37]) and $A = \sqrt{q/\pi}$. We also recall that for quadratic characters χ, the root number ϖ is 1. We logarithmically differentiate this expression to obtain:

$$\log A + \frac{1}{2}\psi((s+1)/2) + \frac{L'}{L}(s,\chi) = -\log A - \frac{1}{2}\psi((2-s)/2) - \frac{L'}{L}(1-s,\overline{\chi}),$$

where $\psi(s)$ denotes the digamma function, which is the logarithmic derivative of the gamma function. Putting $s = 1$ into the formula, and using (see, for example, p. 301 of [99])

$$\psi(1) = -\gamma, \quad \psi(1/2) = -\gamma - 2\log 2$$

we deduce

$$\frac{L'}{L}(1,\chi) = -2\log A + \gamma + \log 2 - \frac{L'}{L}(0,\overline{\chi}). \qquad (26.8)$$

Now we specialise our discussion to quadratic characters associated with an imaginary quadratic field K. Such a character is necessarily odd and if K has discriminant $-D$, then this character, which we denote by χ_D is a primitive character modulo D. In this situation, we have from the functional equation

$$L(0, \chi_D) = 2h_D/w_D, \tag{26.9}$$

where h_D and w_D denote the class number and number of roots of unity of K. Thus, injecting formula (26.7) into (26.8), we get on exponentiating,

$$\exp\left(\frac{L'(1, \chi_D)}{L(1, \chi_D)} - \gamma\right) = (2D/A^2) \prod_{a=1}^{D} \Gamma(a/D)^{-\chi_D(a)w_D/2h_D}.$$

By Proposition 26.14 and the fact that $A = \sqrt{D/\pi}$, Theorem 26.15 is now immediate. $\square$

Thus we have from the above theorem that

$$\exp\left(\frac{L'(1, \chi_D)}{L(1, \chi_D)} - \gamma\right)$$

is transcendental. In particular,

$$\frac{L'(1, \chi_D)}{L(1, \chi_D)} \neq \gamma,$$

for any D. More generally, we have:

Corollary 26.16

$$\frac{L'(1, \chi_D)}{L(1, \chi_D)} - \gamma$$

is not equal to logarithm of an algebraic number.

From the theorem, we can also deduce the following curious corollary.

Corollary 26.17 *If for some D, we have $L'(1, \chi_D) = 0$, then e^γ is transcendental.*

It is unlikely that such a D exists for a variety of reasons. But this seems difficult to prove. We shall have the occasion to come back to this theme at the end of this chapter.

Theorem 26.15 allows one to connect this non-vanishing question to special values of the Γ-function via the Chowla–Selberg formula. Indeed, the proof of Theorem 26.15 leads to a simple proof of the Chowla–Selberg formula which we spell out now. We can combine the calculations done previously with Corollary 26.12 to deduce the Chowla–Selberg formula:

$$\prod_{\mathfrak{C} \in \mathcal{H}_K} g(\mathfrak{C})^{1/3} = \left(\frac{1}{2\pi D}\right)^{2h_D} \prod_{a=1}^{D} \Gamma(a/D)^{w_D \chi_D(a)}.$$

Let us analyse the left-hand side of this equation following [55]. Let E be an elliptic curve with complex multiplication by an order in the imaginary quadratic field $K = \mathbb{Q}(\sqrt{-D})$. Formula (3) of [55] states that any period of E, up to an algebraic factor, is given by the right-hand side of the above equation. In other words,

$$f(\chi_D) := \prod_{a=1}^{D} \Gamma(a/D)^{\chi_D(a)}$$

is equal to a product of a power of π and a power of the period of the CM elliptic curve attached to the full ring of integers of $\mathbb{Q}(\sqrt{-D})$ (up to an algebraic factor).

More generally, we can define for any character $\chi \pmod D$,

$$f(\chi) = \prod_{a=1}^{D} \Gamma(a/D)^{\chi(a)}.$$

Let $q \mid D$ and χ be a real primitive character $\pmod q$. Let χ^* denote the character obtained by extending χ to residue classes $\pmod D$. Then, it is not hard to see that $f(\chi^*)$ is (up to a non-zero algebraic factor) equal to $f(\chi)$. Indeed, recall that

$$\Gamma(z)\Gamma(z + 1/q) \cdots \Gamma(z + (q-1)/q) = q^{1/2 - qz}(2\pi)^{(q-1)/2}\Gamma(qz).$$

Thus, $f(\chi^*) =$

$$\prod_{a=1}^{D} \Gamma(a/D)^{\chi^*(a)} = \prod_{a=1}^{q} [\Gamma(a/D)\Gamma((a+q)/D) \cdots \Gamma((a + (D/q - 1)q)/D)]^{\chi(a)}$$

$$= \prod_{a=1}^{q} [\Gamma(a/q)(2\pi)^{(D/q-1)/2}(D/q)^{1/2 - a/q}]^{\chi(a)}$$

$$= f(\chi)(D/q)^{\sum_{a=1}^{q}(1/2 - a/q)\chi(a)}.$$

Since χ is a real character, the exponent of D/q is rational and so the second factor is algebraic and non-zero. Consequently, $f(\chi)$ and $f(\chi^*)$ are equal apart from a non-zero algebraic factor. Let us record this remark here since it will be used later.

One can consider a more general situation where one considers functions f defined on ray class groups and similar formulas and results can be derived (see [98], for instance).

These investigations naturally lead to the study of possible transcendence of special values of the Γ-function. As we have seen before, not much is known in this context. While $\Gamma(1/2) = \sqrt{\pi}$ is transcendental by the theorem of Lindemann, the transcendence of $\Gamma(1/3)$ and $\Gamma(1/4)$ was established by Chudnovsky [32] in 1976. Recently, Grinspan [53] and Vasilev [124] independently showed that at least two of the three numbers $\pi, \Gamma(1/5), \Gamma(2/5)$ are algebraically independent. Very likely, all of the three numbers are algebraically independent.

Apart from these results, no further results are known regarding the transcendence of the Γ-function at rational, non-integral arguments. Thus, in this context, the following theorem is of interest.

Theorem 26.18 *Let $q > 1$ and $q|24$. Let a be coprime to q. There exists a finite set S and a collection of pair-wise non-isogenous CM elliptic curves E_j, $j \in S$ defined over $\overline{\mathbb{Q}}$ with fundamental real periods ω_j such that $\Gamma(a/q)$ lies in the field generated over $\mathbb{Q}$ by π and the ω_j. In particular, if π and the ω_j's are algebraically independent, then $\Gamma(a/q)$ is transcendental.*

Proof. For each odd quadratic character χ_D, we have an associated imaginary quadratic extension k_D. Thus, $f(\chi)$ is defined for any odd quadratic character. We can associate a CM elliptic curve E_D, with ring of endomorphisms isomorphic to the ring of integers of k_D. Let ω_D be the real period of E_D. The Chowla–Selberg formula expresses

$$\sum_{a=1}^{D} \chi_D(a) \log \Gamma(a/D)$$

as a $\mathbb{Q}$-linear form in $\log \pi$, $\log \omega_D$ and the logarithm of a non-zero algebraic number. For any divisor q of 24, every non-trivial Dirichlet character mod q is quadratic. Noting that

$$\sum_{\chi \text{ even}} \chi(a) = \varphi(q)/2,$$

if $a \equiv \pm 1 (\bmod\, q)$ and zero otherwise, we deduce that

$$\sum_{\chi \text{ odd}} \chi(a) = \begin{cases} \varphi(q)/2 & \text{if } a \equiv 1 (\bmod\, q) \\ -\varphi(q)/2 & \text{if } a \equiv -1 (\bmod\, q) \\ 0 & \text{otherwise.} \end{cases}$$

By considering the combination

$$\prod_{\chi \text{ odd}} f(\chi)^{\chi(b)}$$

where the product is over odd characters $(\bmod\, q)$, we find

$$\prod_{\chi \text{ odd}} f(\chi)^{\chi(b)} = \prod_{a=1}^{q} \Gamma(a/q)^{\sum_{\chi \text{ odd}} \chi(ab)}.$$

Since for any divisor q of 24, $b^2 \equiv 1(\bmod\, q)$ for any b coprime to q, we have $ab \equiv 1(\bmod\, q)$ implies $a \equiv b(\bmod\, q)$. Thus,

$$\sum_{\chi \text{ odd}} \chi(ab) = \begin{cases} \varphi(q)/2 & \text{if } a \equiv b(\bmod\, q) \\ -\varphi(q)/2 & \text{if } a \equiv -b(\bmod\, q) \\ 0 & \text{otherwise.} \end{cases}$$

and we deduce that

$$\Gamma(a/q)\Gamma(1 - a/q)^{-1}$$

is the product of an algebraic number, a power of π and a product of powers of periods of non-isogenous elliptic curves. On the other hand,

$$\Gamma(a/q)\Gamma(1 - a/q)$$

is a product of π and an algebraic number. Thus, we deduce that $\Gamma(a/q)$ is (up to an algebraic factor) a product of a power of π and periods of non-isogenous elliptic curves. This completes the proof. $\square$

To summarise, the key point here is that the non-trivial Dirichlet characters (mod 24) are all quadratic which allows the use the Chowla–Selberg formula as stated before to express $\Gamma(a/q)$ as a product of π and periods of various non-isogenous elliptic curves.

Before moving on in our discussion, we observe this amusing corollary of the above theorem:

Corollary 26.19 *All of the numbers*

$$\Gamma(1/8), \Gamma(3/8), \Gamma(5/8), \Gamma(7/8)$$

are transcendental with at most one exception.

Proof. To prove this, we suppose that at least two of the numbers, $\Gamma(a/8)$, $\Gamma(b/8)$ (say), among

$$\Gamma(1/8), \Gamma(3/8), \Gamma(5/8), \Gamma(7/8)$$

are algebraic. By the proof of the previous theorem, we can write each term as a product of powers of π and periods ω_1 and ω_2 of two non-isogenous CM elliptic curves. By taking appropriate powers of $\Gamma(a/8), \Gamma(b/8)$, we deduce that their quotient, which is algebraic, is a product of powers of π and ω_1. By the result of Chudnovsky [32], we know that π and ω_1 are algebraically independent. This completes the proof. $\square$

Recall that Schanuel's conjecture predicts that if $x_1, \ldots, x_n$ are linearly independent over $\mathbb{Q}$, then the transcendence degree of the field

$$\mathbb{Q}(x_1, \ldots, x_n, e^{x_1}, \ldots, e^{x_n})$$

is at least n. At the end of Chap. 21, an elliptic-exponential extension of this conjecture has been spelt out. The previous theorem motivates the following variant of Schanuel's conjecture.

Suppose that $x_1, \ldots, x_n$ are linearly independent over $\overline{\mathbb{Q}}$. Let $\wp_2, \ldots, \wp_n$ be the Weierstrass $\wp$-functions attached to non-isogenous CM elliptic curves

$E_2, \ldots, E_n$ defined over $\overline{\mathbb{Q}}$. If $x_2, \ldots, x_n$ are not contained in the poles of the $\wp_i$, $2 \leq i \leq n$, then, the transcendence degree of the field

$$\mathbb{Q}(x_1, \ldots, x_n, e^{x_1}, \wp_2(x_2), \ldots, \wp_n(x_n))$$

is at least n.

Thus, choosing $x_1 = \pi i$ and $x_j = \omega_j/2$ with the ω_j as in Theorem 26.18, the conjecture allows us to deduce that π and the ω_j's are algebraically independent.

The above conjecture is also a special case of a more general conjecture of Grothendieck (see [39]). This conjecture asserts that the transcendence degree of the field generated by the periods of an algebraic variety is equal to d where d is the dimension of the Hodge group of the variety. In our case, we consider the variety

$$X = \mathbb{P}^1 \times E_2 \times \cdots \times E_n$$

where E_i are pairwise non-isogenous elliptic curves with complex multiplication. The Hodge group of $H^2(\mathbb{P}^1) \otimes \cdots \otimes H^1(E_n)$ is isomorphic to

$$\mathbb{G}_m \times \prod_{i=2}^{n} (R_{K_i/\mathbb{Q}} \mathbb{G}_m)^1,$$

where K_i is the imaginary quadratic field corresponding to E_i and the superscript denotes elements of norm 1. The dimension of this group is n.

It is clear from the preceding discussions that the non-vanishing of certain Dirichlet series is connected with linear independence of special values of L-series. Such a theme was explored in a classical context in [61].

Also it highlights the pivotal role played by $L'(1, \chi)$ with χ a Dirichlet character, more precisely the vanishing of $L'(1, \chi)$ for any Dirichlet character χ (mod q). In this context, we shall derive an analytic result about the number of $\chi \neq \chi_0$ (mod q) for which $L'(1, \chi) = 0$

For this we shall use the following result of Y. Ihara, K. Murty and M. Shimura which is Theorem 5 of [68].

Proposition 26.20 *Let $\Lambda_0(1) = 1$ and $\Lambda_0(n) = 0$ for $n > 1$. Define for $k \geq 1$,*

$$\Lambda_k(n) = \sum_{n_1 \cdots n_k = n} \Lambda(n_1) \cdots \Lambda(n_k),$$

where Λ denotes the von Mangoldt function. Set

$$\mu^{(a,b)} := \sum_{n=1}^{\infty} \frac{\Lambda_a(n) \Lambda_b(n)}{n^2}.$$

Then, for q prime and any $\epsilon > 0$

$$T_{a,b} := \sum_{\chi \neq \chi_0} P^{(a,b)} \left(\frac{L'}{L}(1, \chi) \right) = (-1)^{a+b} \mu^{(a,b)} \varphi(q) + O(q^\epsilon),$$

where $P^{(a,b)}(z) = z^a \overline{z}^b$.

It is easy to see that the series for $\mu^{(a,b)}$ converges. Indeed, $\Lambda(n) \leq \log n$ so that $\Lambda_k(n) \leq d_k(n)(\log n)^k$, where $d_k(n)$ denotes the number of factorisations of n as a product of k natural numbers. Consequently, $\Lambda_k(n) = O(n^\epsilon)$ for any $\epsilon > 0$.

We now have the following theorem:

Theorem 26.21 *For q prime, the number of $\chi \neq \chi_0$ (mod q) for which $L'(1,\chi) = 0$ is $O(q^\epsilon)$ for any $\epsilon > 0$.*

Proof. We apply the previous proposition with $a = b = k$ and $a = b = 2k$. An application of the Cauchy–Schwarz inequality to the sum

$$\sum_{\chi \neq \chi_0} P^{(k,k)} \left(\frac{L'}{L}(1,\chi) \right)$$

shows that for any $k \geq 1$,

$$\#\{\chi \neq \chi_0 : L'(1,\chi) \neq 0\} \geq \frac{T_{k,k}^2}{T_{2k,2k}}.$$

Let us note that

$$T_{k,k}^2 = (\mu^{(k,k)})^2 \varphi(q)^2 + O(\varphi(q)q^\epsilon)$$

and that

$$(\mu^{(k,k)})^2 = \sum_{n_1,n_2} \frac{\Lambda_k(n_1)^2 \Lambda_k(n_2)^2}{n_1^2 n_2^2} = \sum_{n=1}^{\infty} \frac{(\Lambda_k^2 \star \Lambda_k^2)(n)}{n^2},$$

where

$$(f \star g)(n) := \sum_{d|n} f(d)g(n/d),$$

is the Dirichlet convolution. Now, if $d(n)$ denotes the number of divisors of n,

$$\Lambda_{2k}(n)^2 = (\Lambda_k \star \Lambda_k)^2(n) = \left(\sum_{d|n} \Lambda_k(d)\Lambda_k(n/d) \right)^2$$

$$\leq d(n) \sum_{d|n} \Lambda_k^2(d)\Lambda_k^2(n/d) = d(n) \left(\Lambda_k^2 \star \Lambda_k^2 \right)(n),$$

by an application of the Cauchy–Schwarz inequality. As $d(n) = O(n^\epsilon)$ for any $\epsilon > 0$, we obtain

$$\mu^{(2k,2k)} = \sum_{n=1}^{\infty} \frac{\Lambda_{2k}^2(n)}{n^2} \leq \sum_{n=1}^{\infty} \frac{(\Lambda_k^2 \star \Lambda_k^2)(n)}{n^{2-\epsilon}}.$$

Putting

$$G_k(s) = \sum_{n=1}^{\infty} \frac{(\Lambda_k^2 \star \Lambda_k^2)(n)}{n^s}$$

we conclude

$$\frac{T_{k,k}^2}{T_{2k,2k}} \geq \frac{G_k(2)\varphi(q)^2 + O(\varphi(q)q^{\epsilon_2})}{G_k(2-\epsilon_1)\varphi(q) + O(q^{\epsilon_2})}$$

for any $\epsilon_1, \epsilon_2 > 0$. Choosing $k = 2$ and noting that

$$G_2(2 - \epsilon_1) = G_2(2) + O(\epsilon_1),$$

we conclude that

$$\frac{T_{k,k}^2}{T_{2k,2k}} \geq \varphi(q) + O(q^\epsilon).$$

The result immediately follows from choosing $\epsilon_1 = 1/q$. $\square$

It is unlikely that one can show the non-vanishing of $L'(1,\chi)$ in general using such analytic methods.

The question of non-vanishing of $L'(1,\chi)$ arises in other contexts like the following. Let K be an algebraic number field and $\zeta_K(s)$ its Dedekind zeta function. It is well known that $\zeta_K(s)$ has a simple pole at $s = 1$ with residue λ_K. Here,

$$\lambda_K = \frac{2^{r_1}(2\pi)^{r_2} h_K R_K}{w\sqrt{|d_K|}},$$

where r_1 is the number of real embeddings of K and $2r_2$ is the number of non-real embeddings of K, h_K, R_K, w and d_K are the class number, regulator, the number of units of finite order and discriminant, respectively, of K. Let us set

$$g_K(s) = \zeta_K(s) - \lambda_K \zeta(s).$$

Then, $g_K(s)$ is analytic at $s = 1$. In [111], Scourfield asked if for any field $K \neq \mathbb{Q}$ we have $g_K(1) = 0$. This question is really about non-vanishing of linear combinations of derivatives of L-functions.

To see this, we write

$$\zeta_K(s) = \zeta(s)F_K(s),$$

where $F_K(s)$ is a product of certain Artin L-series. Using Brauer's induction theorem and the non-vanishing of Hecke L-series at $s = 1$, it is easily seen that $F_K(s)$ is analytic at $s = 1$. Consequently, $F_K(1) = \lambda_K$ and since

$$\zeta_K(s) - \lambda_K \zeta(s) = \zeta(s)(F_K(s) - \lambda_K),$$

we see that $g_K(1) = F_K'(1)$. If $\hat{K}$ denotes the normal closure of K over $\mathbb{Q}$, and $G = \mathrm{Gal}(\hat{K}/\mathbb{Q})$, one can express $F_K(s)$ as a product of Artin L-series attached to irreducible characters of G. Indeed, if $H = \mathrm{Gal}(\hat{K}/K)$, $\zeta_K(s)$ is the Artin L-series attached to the character $\mathrm{Ind}_H^G 1$. If χ is an irreducible character of G, we have by Frobenius reciprocity,

$$c_\chi := (\mathrm{Ind}_H^G 1, \chi) = (1, \chi|_H)$$

which is the multiplicity of the trivial character in χ restricted to H. Thus, c_χ is a non-negative integer and we have

$$F_K(s) = \prod_{\chi \neq 1} L(s, \chi)^{c_\chi},$$

where the product is over the non-trivial irreducible characters of G. Hence,

$$\frac{F_K'(1)}{F_K(1)} = \sum_{\chi \neq 1} c_\chi \frac{L'}{L}(1, \chi).$$

In the special case $K/\mathbb{Q}$ is Galois, $c_\chi = \chi(1)$. Thus, in the Galois case, the question of non-vanishing of $g_K(1)$ is equivalent to the non-vanishing of

$$\sum_{\chi \neq 1} \chi(1) \frac{L'}{L}(1, \chi).$$

If $K = \mathbb{Q}(\zeta_q)$ is the q-th cyclotomic field, with ζ_q being a primitive q-th root of unity, then Ihara et al. [68] have investigated the asymptotic behaviour of this sum. They proved that

$$\lim_{q \to \infty, q \text{ prime}} \frac{1}{\phi(q)} \sum_{\chi \neq \chi_1} \frac{L'}{L}(1, \chi) = 0.$$

So the question of non-vanishing of $g_K(1)$ is a bit delicate and cannot be deduced from this limit theorem.

The non-vanishing of $L'(1, \chi)$ seems to be intimately linked with arithmetic questions. For example, if $K/\mathbb{Q}$ is quadratic, then $F_K(s) = L(s, \chi_D)$ where χ_D is the quadratic character attached to K. In this case, Scourfield's question reduces to the question of whether $L'(1, \chi_D) = 0$ for any such χ_D. As we mentioned before, it is unlikely that such a χ_D exists.

Exercises

1. Using (26.2), show that the series (26.1) admits a meromorphic continuation to $\Re(s) > \frac{d}{d+1}$ with at most a simple pole at $s = 1$. Moreover, the simple pole exists if and only if $\rho_f \neq 0$.

2. Show that the character χ_{D_1, D_2} given by (26.6) is well defined.

3. Prove that $\zeta'(0, x) = \log(\Gamma(x)/2\pi)$.

4. Prove the class number formula (26.9).

5. Show that the function

$$g(\mathfrak{b}) = (2\pi)^{-12} (\mathbf{N}(\mathfrak{b}))^6 \Delta(\beta_1, \beta_2)$$

considered in the chapter is well defined and does not depend on the choice of integral basis of $\mathfrak{b}$. Furthermore, show that $g(\mathfrak{b})$ depends only on the ideal class $\mathfrak{b}$ belongs to in the ideal class group.

6. Show that $L(1,\chi) = L(1,\overline{\chi})$ for any non-trivial ideal class character χ of an imaginary quadratic field.

7. For any integer a, find the value of

$$\sum_{\chi \text{ even}} \chi(a)$$

where χ runs over even Dirichlet characters mod q. Derive also an expression for

$$\sum_{\chi \text{ odd}} \chi(a)$$

where χ runs over odd Dirichlet characters mod q.

Chapter 27

Transcendence of Values of Modular Forms

In this chapter, we will apply the results of Schneider and Nesterenko to investigate the values of modular forms at algebraic arguments. Any reasonable account of the fascinating subject of modular forms will require us to embark upon a different journey which we cannot undertake in the present book. We refer to the books [42, 70, 75] for comprehensive accounts of this subject. However for the purposes of this chapter, we shall be needing very little input from the theory of modular forms.

As we have been observing throughout, the naturally occurring transcendental functions like the exponential function and the logarithm function take transcendental values when evaluated at algebraic points, except for some obvious exceptions. This is also exhibited by the Weierstrass-$\wp$ function associated with an elliptic curve defined over number fields. We also expect other transcendental functions like the gamma function and Riemann zeta function to exhibit similar properties.

We will now investigate this phenomena for modular forms which are a rich source of transcendental functions. We begin by fixing some notations and recalling the various results in transcendence relevant for our study.

Let $\mathbb{H}$ denote the upper half-plane. For $z \in \mathbb{H}$, we have the following functions

$$E_2(z) = 1 - 24 \sum_{n=1}^{\infty} \sigma_1(n) e^{2\pi i n z},$$

M.R. Murty and P. Rath, *Transcendental Numbers*, DOI 10.1007/978-1-4939-0832-5_27, 179
© Springer Science+Business Media New York 2014

$$E_4(z) = 1 + 240 \sum_{n=1}^{\infty} \sigma_3(n) e^{2\pi i n z},$$

$$E_6(z) = 1 - 504 \sum_{n=1}^{\infty} \sigma_5(n) e^{2\pi i n z},$$

where $\sigma_k(n) = \sum_{d|n} d^k$. We also have the j-function given by

$$j(z) = 1728 \ \frac{E_4(z)^3}{E_4(z)^3 - E_6(z)^2} \ .$$

We call an element α in the upper half-plane to be a CM point if it generates a quadratic extension over the field of rational numbers. It is known, from classical theory of complex multiplication, that if $z \in \mathbb{H}$ is a CM point, then $j(z)$ is an algebraic number lying in the Hilbert class field of $\mathbb{Q}(z)$. For instance, we have $j(i) = 1728$ while $j(\rho) = 0$ where $\rho = e^{2\pi i/3}$.

For algebraic points in the upper half-plane, we have already seen the following result of Schneider:

Theorem 27.1 (Schneider) *If $z \in \mathbb{H}$ is algebraic, then $j(z)$ is algebraic if and only if z is CM.*

Much later, Chudnovsky ([32], see also [33]) in 1976 showed that if $z \in \mathbb{H}$, then at least two of the numbers $E_2(z)$, $E_4(z)$, $E_6(z)$ are algebraically independent. Chudnovsky's theorem proves that $\Gamma(1/3)$ and $\Gamma(1/4)$ are transcendental. In 1995, Barré-Sirieix et al. [15] made a breakthrough in transcendence theory by proving the long-standing conjecture of Mahler and Manin according to which the modular invariant $J(e^{2\pi i z}) := j(z)$ assumes transcendental values at any non-zero complex (or p-adic) algebraic number $e^{2\pi i z}$ in the unit disc. Note that such a z is necessarily transcendental. Finally, Nesterenko [84] provided a fundamental advance by generalising both the results of Chudnovsky and Barré-Sirieix–Diaz–Gramain–Philibert.

Theorem 27.2 (Nesterenko) *Let z be a point in the upper half-plane. Then at least three of the four numbers*

$$e^{2\pi i z}, \quad E_2(z), \quad E_4(z), \quad E_6(z)$$

are algebraically independent.

We note that the result of Schneider does not follow from the theorem of Nesterenko. As pointed out by Nesterenko [86, p. 31], both his and Schneider's theorem will follow from the following conjecture:

Conjecture 27.3 *Let z be a point in the upper half-plane and assume that at most three of the following five numbers*

$$z, \quad e^{2\pi i z}, \quad E_2(z), \quad E_4(z), \quad E_6(z)$$

are algebraically independent. Then z is necessarily a CM *point and the field*

$$\mathbb{Q}(e^{2\pi i z}, E_2(z), E_4(z), E_6(z))$$

has transcendence degree 3.

Let us now begin by considering the nature of zeros of modular forms. Investigations of such zeros have been carried out by Rankin and Swinnerton-Dyer [104], Kanou [69], Kohnen [73] and Gun [57] (see also [13, 45, 49]). Let us again recall that a CM point is an element of $\mathbb{H}$ lying in an imaginary quadratic field. Also every modular form is assumed to be non-zero and for the full modular group. However, the arguments carry over to higher levels.

Recall that any such modular form f has a q-expansion at infinity of the form $f(z) = \sum_{n=0}^{\infty} a_f(n)e^{2\pi i n z}$ for $z \in \mathbb{H}$. The $a_f(n)'s$ are called the Fourier coefficients of f. We shall only consider modular forms whose Fourier coefficients are all algebraic.

To study the algebraic nature of values taken by modular forms, we need to define an equivalence relation on the set of all modular forms with algebraic Fourier coefficients. We define two such modular forms f and g to be equivalent, denoted by $f \sim g$, if there exist positive natural numbers k_1, k_2 such that $f^{k_2} = \lambda g^{k_1}$ with $\lambda \in \overline{\mathbb{Q}}^*$. Furthermore, for the purpose of this chapter, we shall denote Δ to be the Ramanujan cusp form (or the normalised discriminant function), i.e.

$$\Delta(z) = q \prod_{n=1}^{\infty} (1 - q^n)^{24} = \eta(z)^{24}, \qquad q = e^{2\pi i z}.$$

Therefore, it is also equal to

$$\Delta(z) = \frac{E_4(z)^3 - E_6(z)^2}{1728}.$$

One has the following theorem.

Theorem 27.4 *Let f be a non-zero modular form of weight k for the full modular group* $SL_2(\mathbb{Z})$*. Suppose that the Fourier coefficients of f are algebraic. Then any zero of f is either* CM *or transcendental.*

Proof. Let f be a non-zero modular form of weight k for $SL_2(\mathbb{Z})$ with algebraic Fourier coefficients. Let g be the function defined as

$$g(z) = \frac{f^{12}(z)}{\Delta^k(z)},$$

where Δ is the Ramanujan cusp form of weight 12. Thus g is a modular function of weight 0 and hence is a rational function in j. Since Δ does not vanish on $\mathbb{H}$, g is a polynomial in j. Further, since f has algebraic Fourier coefficients, $g(z) = P(j(z))$ where $P(x)$ is a polynomial with algebraic coefficients. If α is a zero of f, then $P(j(\alpha)) = 0$ and hence $j(\alpha)$ is algebraic. Thus by Schneider's theorem, α is either CM or transcendental. This completes the proof. $\square$

As before, let Δ be the unique normalised cusp form of weight 12 for the full modular group. Then the above theorem easily extends to the following and hence we skip the proof.

Theorem 27.5 *Let f be as in the above theorem, not equivalent to Δ and $\alpha \in \mathbb{H}$ be an algebraic number such that $f^{12}(\alpha)/\Delta^k(\alpha)$ is algebraic. Then α is necessarily a* CM *point.*

We note that the above theorem does not say anything about the transcendental zeros of f. However, when f is the Eisenstein series E_k, we have some more information about the location of their zeros. For instance, all the zeros of E_k up to $SL_2(\mathbb{Z})$ equivalence were shown to lie in the arc

$$\{e^{i\theta} \mid \pi/2 \le \theta \le 2\pi/3\}$$

by Rankin and Swinnerton-Dyer [104].

It is worthwhile to point out that for cusp forms, the situation is rather different. Here we have a result due to Rudnick [108] which is as follows: let $\{f_k\}$ be a sequence of L^2-normalised holomorphic cusp forms for $SL_2(\mathbb{Z})$ such that f_k is of weight k, the order of vanishing of f_k at the cusp is $o(k)$ and the masses $y^k |f_k(z)|^2 dV(z)$ (where $dV(z)$ stands for the normalised hyperbolic measure on the fundamental domain) tend in the weak star topology to $c\, dV(z)$ for some constant $c > 0$. Then the zeros of f_k (in the fundamental domain) are equidistributed with respect to $dV(z)$. If the sequence f_k consists of normalised Hecke eigenforms ordered by increasing weight, then recent works of Soundararajan [119] and Holowinsky [65] (see also [66]) show that Rudnick's hypothesis is satisfied and consequently, zeros of normalised Hecke eigenforms become uniformly distributed in the standard fundamental domain as the weight tends to infinity.

If f is equivalent to Δ and α is CM, then $f(\alpha)$ is transcendental by the theorem of Schneider. On the other hand, if $\alpha \in \mathbb{H}$ is non-CM algebraic, the conjecture of Nesterenko mentioned before will imply the transcendence of $f(\alpha)$. Thus, it is clear that while investigating the nature of values of modular forms at algebraic numbers in $\mathbb{H}$, we need to consider the values at CM points and non-CM points separately.

Theorem 27.6 *Let $\alpha \in \mathbb{H}$ be such that $j(\alpha) \in \overline{\mathbb{Q}}$. Then $e^{2\pi i\alpha}$ and $\Delta(\alpha)$ are algebraically independent.*

Proof. Since $j(\alpha)$ is algebraic, $\Delta(\alpha)$ is transcendental. For, algebraicity of $\Delta(\alpha)$ will imply that $j(\alpha)\Delta(\alpha) = E_4(\alpha)^3$ is algebraic and hence both $E_4(\alpha)$ and $E_6(\alpha)$ are algebraic. This will contradict Chudnovsky's theorem. Now suppose that $e^{2\pi i\alpha} = q$ and $\Delta(\alpha)$ are algebraically dependent. Since $\Delta(\alpha)$ is transcendental, there exists a non-zero polynomial $P(X) = \sum_i p_i X^i$ where

p_i's are polynomials in $\Delta(\alpha)$ with algebraic coefficients such that $P(q) = 0$. Thus q is algebraic over the field $\overline{\mathbb{Q}}(E_4(\alpha), E_6(\alpha))$. Since $j(\alpha)$ is algebraic, transcendence degree of $\overline{\mathbb{Q}}(E_4(\alpha), E_6(\alpha))$ is one which is also the transcendence degree of $\overline{\mathbb{Q}}(E_4(\alpha), E_6(\alpha), q)$. This will contradict Nesterenko's theorem. $\square$

As a consequence of the above theorem, we now have the following:

Theorem 27.7 *Let $\alpha \in \mathbb{H}$ be such that $j(\alpha) \in \overline{\mathbb{Q}}$. Then for a non-zero modular form f for $\mathrm{SL}_2(\mathbb{Z})$ with algebraic Fourier coefficients, $f(\alpha)$ is algebraically independent with $e^{2\pi i \alpha}$ except when $f(\alpha) = 0$.*

Proof. Suppose that $f(\alpha)$ is not equal to zero. Since the non-zero number $f^{12}(\alpha)/\Delta^k(\alpha)$ is a polynomial in $j(\alpha)$ with algebraic coefficients, it is algebraic. Thus the fields $\overline{\mathbb{Q}}(q, f(\alpha))$ and $\overline{\mathbb{Q}}(q, \Delta(\alpha))$ have the same transcendence degree and hence the theorem follows from the previous theorem. $\square$

We note that there exist transcendental numbers α for which $j(\alpha)$ is algebraic. This is a consequence of CM theory and surjectivity of the j function. As mentioned before, an algebraic α for which $j(\alpha)$ is algebraic is a CM point. In this case, $\Delta(\alpha)$ can be explicitly expressed as a power of period of an elliptic curve defined over $\overline{\mathbb{Q}}$.

For a non-CM algebraic number, one has the following theorem:

Theorem 27.8 *For $\alpha \in \mathbb{H}$, let S_α be the set of all non-zero modular forms f of arbitrary weight for $\mathrm{SL}_2(\mathbb{Z})$ with algebraic Fourier coefficients such that $f(\alpha)$ is algebraic. If $\alpha \in \mathbb{H}$ is a non-CM algebraic number, then S_α has at most one element up to equivalence.*

Proof. Let f and g be modular forms in S_α of weight k_1 and k_2, respectively, where α is a non-CM algebraic number in $\mathbb{H}$. Thus both $f(\alpha)$ and $g(\alpha)$ are algebraic and by Theorem 27.4, neither is equal to zero. We consider the modular form $F = f^{k_2}(\alpha)g^{k_1} - g^{k_1}(\alpha)f^{k_2}$ of weight $k_1 k_2$. By Theorem 27.4, any zero of this modular form is either CM or transcendental. Since α is non-CM and algebraic, we get a contradiction unless F is identically zero. This means that f and g are equivalent in the sense defined before. $\square$

The existence of the fugitive exceptional class in the above theorem can be ruled out if we assume the conjecture of Nesterenko alluded to before. Further, all these theorems extend to higher levels and also to quasi-modular forms. We refer to [59] for further details.

Exercises

1. Show that $E_6(\sqrt{-1}) = 0$. Deduce that $e^\pi, E_2(\sqrt{-1})$ and $E_4(\sqrt{-1})$ are algebraically independent.

2. Recalling that

$$E_2\left(\frac{az+b}{cz+d}\right) = (cz+d)^2 E_2(z) + \frac{6c(cz+d)}{\pi i},$$

show that $E_2(\sqrt{-1}) = 3/\pi$. Deduce from the previous exercise that π and e^π are algebraically independent.

3. Show that

$$\omega := \int_1^\infty \frac{dt}{\sqrt{t^3 - t}} = \frac{\Gamma(1/4)^2}{2\sqrt{2\pi}}.$$

Using the theory of complex multiplication, compute the value of $E_4(\sqrt{-1})$ in terms of powers of $\Gamma(1/4)$ and π. Conclude that π, e^π and $\Gamma(1/4)$ are algebraically independent.

4. Let f be a non-zero modular form with algebraic Fourier coefficients and α be an algebraic number in $\mathbb{H}$ which is not CM. Assuming Nesterenko's conjecture, show that $f(\alpha)$ is transcendental.

5. Prove that the functions E_4 and E_6 are algebraically independent.

6. Using the Schneider–Lang theorem, prove that $\Delta(\tau)$ is transcendental for any CM point τ. Here Δ is the normalised discriminant function considered in this chapter.

Chapter 28

Periods, Multiple Zeta Functions and $\zeta(3)$

In this chapter, we will examine some of the emerging themes in the theory of transcendental numbers. The most fascinating is the "modular connection" linking it with the theory of modular forms. We have met a part of this connection in the earlier chapters. In this chapter, we will indicate some other relations.

The connection between the theory of modular forms and transcendental number theory goes back at least a century with the advent of the theory of complex multiplication. More recently there are several major contributions, notably by Nesterenko, relating these two themes of number theory. In addition to this, there is the mysterious proof of Roger Apéry [3] showing the irrationality of $\zeta(3)$ which has been somewhat "explained" by Beukers [20] using the theory of modular forms. However, it is not easy to find an exposition of this theme at the graduate student level or even the senior undergraduate level. It is the purpose of this chapter to highlight this theme and to bring out the salient features of the subject for further study. This chapter is self-contained and can be read independent of the other chapters.

The Algebra of Periods

To keep this chapter self-contained, let us recall that a complex number α is said to be *algebraic* if it satisfies a non-zero polynomial equation with rational coefficients. Otherwise, we say the number is *transcendental*. It was not until

M.R. Murty and P. Rath, *Transcendental Numbers*, DOI 10.1007/978-1-4939-0832-5_28, 185
© Springer Science+Business Media New York 2014

1851, when Liouville using a clever approximation argument managed to give explicit constructions of transcendental numbers. For instance, he showed that

$$\sum_{n=0}^{\infty} \frac{1}{10^{n!}}$$

is transcendental. In 1873, Georg Cantor showed that the algebraic numbers are countable and the real numbers are uncountable. Thus the set of transcendental numbers is uncountable. But deciding whether a given number is transcendental is often a very difficult question. For instance, e was shown to be transcendental by Charles Hermite in 1873 and π was proved transcendental by Lindemann in 1882. In view of the 1734 result proved by Euler, this means that the special values of the Riemann zeta function at even natural numbers are transcendental since

$$2\zeta(2k) = 2\sum_{n=1}^{\infty} \frac{1}{n^{2k}} = -\frac{B_{2k}}{(2k)!}(2\pi i)^{2k}.$$

A similar result is not known for odd values of the Riemann zeta function. In 1978, Apéry surprised the mathematical community by presenting a mysterious proof that $\zeta(3)$ is irrational. After more than 25 years, we can explain some aspects of his proof using the theory of modular forms, but cannot say we understand the proof completely or why it worked. The purpose of this chapter is to explore this theme in some detail and also present it from the context of the theory of periods.

A *period*, as defined by Kontsevich and Zagier [74], is a complex number whose real and imaginary parts are values of absolutely convergent integrals of rational functions with rational coefficients over domains in $\mathbb{R}^n$ given by polynomial inequalities with rational coefficients. The set of periods is denoted by $\mathcal{P}$.

In the above definition, we can replace rational functions and rational coefficients by algebraic functions and algebraic coefficients, respectively, without changing the original set of periods $\mathcal{P}$. This is because we can introduce more variables into the integration process.[1]

$\sqrt{2}$ is a period since

$$\sqrt{2} = \int_{2x^2 \leq 1} dx.$$

All algebraic numbers are periods. The simplest transcendental number which is a period is π since

$$\pi = \int_{x^2+y^2 \leq 1} dx dy.$$

---

[1]A more precise definition can be given as follows. Let X be a smooth quasi-projective variety, $Y \subset X$ a subvariety, ω a closed algebraic n-form on X vanishing on Y, and all defined over $\overline{\mathbb{Q}}$. Let C be a singular n-chain on $X(\mathbb{C})$ with boundary contained in $Y(\mathbb{C})$. Then the integral $\int_C \omega$ is called a *period*.

The set of periods $\mathcal{P}$ contains the algebraic numbers and many interesting transcendental numbers like π. Note that the set of periods is countable and hence the set of numbers which are **not** periods is uncountable. However we do not have an explicit number which has been shown to be a non-period. For instance, is e a period? How about Euler's constant γ? Most likely, these numbers are not periods. An open question is if $1/\pi$ is a period.

It is clear that the set of periods forms a ring under addition and multiplication. Again, it is not known whether this ring has any units other than the obvious ones, namely the non-zero algebraic numbers. We recommend the original article of Kontsevich and Zagier [74] as well as the account by Waldschmidt [126] for further details (see also the paper by Ayoub [7] in this context).

An important class of periods is supplied by the special values of the Riemann zeta function and more generally the multiple zeta values.

The Riemann zeta function $\zeta(s)$ is defined for $\Re(s) > 1$ by the Dirichlet series

$$\zeta(s) = \sum_{n=1}^{\infty} \frac{1}{n^s}.$$

More generally, one can define the multiple zeta function,

$$\zeta(s_1, s_2, \ldots, s_r) = \sum_{n_1 > n_2 > \cdots > n_r \geq 1} \frac{1}{n_1^{s_1} n_2^{s_2} \cdots n_r^{s_r}}$$

and study it as a function of the r complex variables $s_1, \ldots, s_r$. Here, we will be concerned with the theory of special values of these functions, or more precisely, the multiple zeta values $\zeta(s_1, \ldots, s_r)$ with $s_1, s_2, \ldots, s_r$ positive integers and $s_1 \geq 2$ in order to ensure convergence.

One can express these as periods, in the sense defined above. For example, we have

$$\zeta(k) = \int_{1 > t_1 > \cdots > t_k > 0} \frac{dt_1}{t_1} \cdots \frac{dt_{k-1}}{t_{k-1}} \frac{dt_k}{1 - t_k},$$

as is easily verified by direct integration. Also,

$$\zeta(2, 1) = \int_{1 > t_1 > t_2 > t_3 > 0} \frac{dt_1}{t_1} \frac{dt_2}{1 - t_2} \frac{dt_3}{1 - t_3}.$$

Similarly, we define inductively the iterated integral of continuous differential forms $\phi_1, \ldots, \phi_m$ on $[a, b]$ as

$$\int_a^b \phi_1 \cdots \phi_m := \int_a^b \phi_1(t) \int_a^t \phi_2 \cdots \phi_m$$

with the convention that the value is 1 when $m = 0$. If we define two differential forms

$$\omega_0 = \frac{dt}{t}, \quad \omega_1 = \frac{dt}{1 - t},$$

then one can easily show that

$$\zeta(s_1,\ldots,s_r) = \int_0^1 \omega_0^{s_1-1}\omega_1 \cdots \omega_0^{s_r-1}\omega_1.$$

By a theorem of Chen in algebraic topology, the product of such integrals is again a linear combination of such integrals given by the "shuffle product".

With respect to the transcendental nature of the multiple zeta values, Zagier [134] has made the following conjecture. Let V_k be the $\mathbb{Q}$-vector space in $\mathbb{R}$ generated by the multiple zeta values $\zeta(s_1,\ldots,s_r)$ with *weight* $s_1+\cdots+s_r = k$. Set $V_0 = \mathbb{Q}$, $V_1 = 0$. Clearly, $V_2 = \mathbb{Q}\pi^2$. Then, using the shuffle relations, we see that

$$V_k V_{k'} \subseteq V_{k+k'}.$$

If we denote by V the $\mathbb{Q}$-vector space generated by all the V_k's, then Goncharov conjectures that

$$V = \oplus_{k=0}^\infty V_k.$$

Zagier predicts that if $d_k = \dim V_k$, then for $k \geq 3$,

$$d_k = d_{k-2} + d_{k-3} \quad \text{with} \quad d_0 = 1, d_1 = 0, d_2 = 1.$$

In other words,

$$\sum_{k=0}^\infty d_k t^k = \frac{1}{1 - t^2 - t^3}.$$

It is generally suspected that this conjecture would imply the algebraic independence of $\pi, \zeta(3), \zeta(5), \ldots$. If we let c_k be the coefficient of t^k of the rational function on the right-hand side of the above conjectural formula, then it is now known by the work of Terasoma [121] as well as the work of Deligne and Goncharov [40] that

$$d_k \leq c_k.$$

Note that while one expects the dimensions d_k of the spaces V_k to grow exponentially in k, we do not have a single example of a space V_k with dimension at least 2. In this context, in a recent work [61], it has been established that a conjecture of Milnor about Hurwitz zeta values implies that infinitely many of these V_k's have dimension at least 2.

We end this section by mentioning a recent result due to Brown [21]. In this work, he proves a conjecture by Hoffman which states that every multiple zeta value is a $\mathbb{Q}$-linear combination of $\zeta(n_1,\ldots,n_r)$ where $n_i \in \{2,3\}$. In particular, Brown's result is a sweeping generalisation of the works of Terasoma, Deligne and Goncharov. An essential ingredient in the proof of Brown was supplied by Zagier [135] which involves a formula for the special multiple zeta values of the form $\zeta(2,\ldots,2,3,2,\ldots,2)$ as rational linear combinations of products $\zeta(m)\pi^{2n}$ with m odd. The works of Terasoma, Deligne, Goncharov as well as Brown involve rather deep algebraic geometry, more precisely the theory of mixed Tate motives. We can do no better than to direct the interested reader to the beautiful Bourbaki talks of Cartier [27] and Deligne [41]. One wonders if there are simpler, more direct proofs of the results of Brown.

Apéry's Proof Revisited

There are many expositions of Apéry's proof of the irrationality of $\zeta(3)$ (see, for example, [123]). We begin by giving a streamlined version of Apéry's proof and then analyse it from the standpoint of modular forms.

Apéry begins by considering the recursion

$$n^3 u_n + (n-1)^3 u_{n-2} = (34n^3 - 51n^2 + 27n - 5)u_{n-1}.$$

Let A_n be the sequence obtained by setting $A_0 = 1, A_1 = 5$ and let B_n be the sequence obtained by setting $B_0 = 0$ and $B_1 = 6$. Thus, if we let $P(n) = 34n^3 - 51n^2 + 27n - 5$, then

$$n^3 A_n + (n-1)^3 A_{n-2} = P(n)A_{n-1},$$

$$n^3 B_n + (n-1)^3 B_{n-2} = P(n)B_{n-1}.$$

Multiplying the first equation by B_{n-1} and the second by A_{n-1} and subtracting, we deduce that

$$n^3(A_n B_{n-1} - A_{n-1} B_n) = (n-1)^3(A_{n-1}B_{n-2} - A_{n-2}B_{n-1}).$$

Iterating, we find

$$n^3(A_n B_{n-1} - A_{n-1} B_n) = A_1 B_0 - A_0 B_1 = -6.$$

Then, Apéry made some remarkable claims. First he asserted that A_n's are all integers. Further, he claimed that B_n's are rational numbers such that

$$2 \operatorname{lcm}[1, 2, 3, \dots, n]^3 B_n$$

are all integers. Finally, one has the following explicit formulas:

$$A_n = \sum_{k=0}^{n} \binom{n}{k}^2 \binom{n+k}{k}^2,$$

$$B_n = \sum_{k=0}^{n} \binom{n}{k}^2 \binom{n+k}{k}^2 c_{n,k}$$

where

$$c_{n,k} = \sum_{j=1}^{n} \frac{1}{j^3} - \sum_{j=1}^{k} \frac{(-1)^j}{2j^3} \binom{n}{j}^{-1} \binom{n+j}{j}^{-1}.$$

It turns out that explicit expressions for the A_n's and the B_n's are not needed to prove the irrationality of $\zeta(3)$. In fact, from the recursion above, we see that

$$\frac{B_n}{A_n} - \frac{B_{n-1}}{A_{n-1}} = \frac{6}{n^3 A_n A_{n-1}}.$$

From the explicit expressions for A_n and B_n, we can deduce that B_n/A_n tends to $\zeta(3)$. Indeed,

$$B_n = A_n \sum_{j=1}^{n} \frac{1}{j^3} - \sum_{k=0}^{n} \sum_{j=1}^{k} \frac{(-1)^j \binom{n}{k}^2 \binom{n+k}{k}^2}{2j^3 \binom{n}{j} \binom{n+j}{j}}.$$

We will show that

$$n^2 \leq j^3 \binom{n}{j} \binom{n+j}{j}.$$

This is clear for $j = n$ since

$$n^3 \binom{2n}{n} \geq n^2.$$

For $1 \leq j \leq n-1$, we have

$$n \leq \binom{n}{j} \leq \binom{n+j}{j}$$

so that

$$n^2 \leq \binom{n}{j} \binom{n+j}{j} \leq j^3 \binom{n}{j} \binom{n+j}{j}.$$

Thus,

$$\left| B_n - A_n \sum_{j=1}^{n} \frac{1}{j^3} \right| \leq \frac{A_n}{2n},$$

from which the assertion that B_n/A_n tends to $\zeta(3)$ follows. One can be a bit more precise. We have by summing

$$\sum_{k=n}^{\infty} \left(\frac{B_{k+1}}{A_{k+1}} - \frac{B_k}{A_k} \right) = 6 \sum_{k=n}^{\infty} \frac{1}{(k+1)^3 A_k A_{k+1}}.$$

$$\zeta(3) - \frac{B_n}{A_n} = 6 \sum_{k=n}^{\infty} \frac{1}{(k+1)^3 A_k A_{k+1}}$$

so that

$$\left| \zeta(3) - \frac{B_n}{A_n} \right| \ll \frac{1}{A_n^2}.$$

We need to estimate the growth of A_n. Again, from the recursion this is easily done. If we divide the recursion by n^3 and take the limit as n tends to infinity, we observe that

$$A_n \sim C_n$$

where C_n satisfies

$$C_n + C_{n-2} = 34 C_{n-1}.$$

The latter recurrence is easily solved and we see that C_n is asymptotically α^n where $\alpha = (1 + \sqrt{2})^4 = 17 + 12\sqrt{2}$ is the larger root of

$$X^2 - 34X + 1 = 0.$$

Finally, we need to prove the assertion about the denominators of B_n. To this end, we observe that

$$\frac{\binom{n}{k}^2 \binom{n+k}{k}^2}{j^3 \binom{n}{j}\binom{n+j}{j}} = \frac{\binom{n}{k}\binom{n+k}{k}\binom{n-j}{n-k}\binom{n+k}{k-j}}{j^3 \binom{k}{j}^2},$$

simply by writing out the binomial coefficients. Thus, we need to investigate the power of a fixed prime p that can appear in the denominator, namely the power of p in

$$j^3 \binom{k}{j}^2.$$

But this is easily done by observing that

$$\mathrm{ord}_p \binom{k}{j} = \sum_{t=0}^{[\log k / \log p]} [k/p^t] - [j/p^t] - [(k-j)/p^t].$$

It is elementary to see that

$$[x + y] - [x] - [y] \leq 1$$

with equality if and only if $\{x\} + \{y\} \geq 1$, where $\{x\}$ denotes the fractional part of x. In particular, we see that the above summation can begin from $t = \mathrm{ord}_p(j) + 1$ which gives us

$$\mathrm{ord}_p \binom{k}{j} \leq [\log k / \log p] - \mathrm{ord}_p(j).$$

Thus,

$$\mathrm{ord}_p \left(j^3 \binom{k}{j}^2 \right) \leq \mathrm{ord}_p(j) + 2[\log k / \log p] \leq 3[\log k / \log p],$$

from which we deduce the statement about denominators. Now suppose that $\zeta(3)$ is rational say C/D, with C, D co prime integers. Then, for the non-zero integer, we have the following estimate (from the Prime Number Theorem):

$$2D \mathrm{lcm}[1, 2, \ldots, n]^3 |A_n \zeta(3) - B_n| \ll e^{3(n + o(n))}(1 + \sqrt{2})^{-4n}.$$

For n large, this is a contradiction because

$$e^3 < (1 + \sqrt{2})^4 = 33.970563 \ldots.$$

In the above proof, the recursion formulas for A_n and B_n were used in two places. We remark that we can eliminate the second use of the recursion where we derived $A_n \sim \alpha^n$. Such an estimate was needed only to get a final upper bound that led to the contradiction. To this end, a simpler estimate suffices if we observe that for n even, we have

$$A_n \geq \binom{n}{n/2}\binom{3n/2}{n/2}.$$

One can see that

$$A_n \geq c_0 2^{5n}/n^4$$

for some constant c_0 and this suffices to get the desired contradiction since

$$e^3 = 20.085537\ldots < 32 = 2^5.$$

Another approach was taken recently by Nesterenko [85]. Following earlier work of Gutnik, he considers the rational function

$$R(z) = \frac{(z-1)^2 \cdots (z-n)^2}{z^2(z+1)^2 \cdots (z+n)^2}.$$

The function R can be expanded into partial fractions:

$$R(z) = \sum_{k=0}^{n}\left(\frac{B_{k2}}{(z+k)^2} + \frac{B_{k1}}{z+k}\right).$$

It is easy to see that

$$B_{k2} = (z+k)^2 R(z)|_{z=-k} = \binom{n+k}{k}^2 \binom{n}{k}^2,$$

and

$$B_{k1} = \frac{d}{dz}((z+k)^2 R(z))|_{z=-k} = -2B_{k2}\left(\sum_{j=1}^{n}\frac{1}{k+j} - \sum_{j=0,j\neq k}^{n}\frac{1}{k-j}\right).$$

It is clear from these expressions that B_{k2} is integral and that $D_n B_{k1}$ is integral where

$$D_n = \prod_{p\leq n} p^{[\log 2n/\log p]}.$$

If we choose a large enough contour C, then the Cauchy residue theorem shows that

$$\frac{1}{2\pi i}\int_C R(z)dz = \sum_{k=0}^{n} B_{k1} = 0,$$

since $R(z) = O(|z|^{-2})$. Using this fact, we deduce that

$$I := \sum_{v=1}^{\infty} R'(v) = \sum_{k=0}^{n} \sum_{v=1}^{\infty} \left(-2\frac{B_{k2}}{(v+k)^3} - \frac{B_{k1}}{(v+k)^2} \right) = a_n\zeta(3) + b_n,$$

where

$$a_n = -2\sum_{k=0}^{n} B_{k2}$$

and

$$b_n = \sum_{k=0}^{n} \sum_{r=1}^{k} \left(2B_{k2}r^{-3} + B_{k1}r^{-2} \right)$$

so that a_n and $D_n^3 b_n$ are integers. On the other hand, one can write down an integral expression for this as

$$I = \frac{1}{2\pi i} \int_{\Re(s)=C} \left(\frac{\pi}{\sin \pi z} \right)^2 R(z)dz.$$

To see this, we truncate the integral from $C + iT$ to $C - iT$ with $T = N + 1/2$, N a positive integer $> n$ tending to infinity. We deform the contour to the rectangle whose vertices are given by $(C, -T), (T, -T), (T, T)$, and (C, T). On the boundary of the rectangle, $1/\sin^2 \pi z$ is bounded and $R(z) = O(1/T^2)$. Thus we compute the contribution from the residues: for $z = k$, k a positive integer, we have

$$\left(\frac{\pi}{\sin \pi z} \right)^2 = \frac{1}{(z-k)^2} + O(1)$$

and

$$R(z) = R(k) + R'(k)(z - k) + O((z - k)^2),$$

so that

$$\text{Res}_{z=k} \left(\left(\frac{\pi}{\sin \pi z} \right)^2 R(z) \right) = R'(k),$$

from which the formula is easily deduced. Finally, using the method of steepest descent, Nesterenko shows that

$$I = \frac{\pi^{3/2} 2^{3/4}}{n^{3/2}} (\sqrt{2} - 1)^{4n+2}(1 + o(1))$$

as n tends to infinity. From this fact, the irrationality of $\zeta(3)$ is easily deduced, as before.

Yet a third proof by Beukers [19] uses the family of polynomials

$$P_n(x) = \frac{1}{n!} \frac{d^n}{dx^n} x^n (1 - x)^n$$

which is easily seen to have integral coefficients. By straightforward integration, it is easy to see that for $r > s$,

$$\int_0^1 \int_0^1 \frac{x^r y^s dx dy}{1 - xy}$$

is a rational number whose denominator is divisible by $d_r^2 = [1, 2, \ldots, r]^2$. Similarly,

$$-\int_0^1 \int_0^1 \frac{x^r y^s \log xy \, dx dy}{1 - xy}$$

is a rational number whose denominator is divisible by d_r^3. If $r = s$, then it is clear that

$$\int_0^1 \int_0^1 \frac{x^r y^r dx dy}{1 - xy} = \zeta(2) - \left(1 + \frac{1}{2^2} + \cdots + \frac{1}{r^2}\right)$$

and

$$-\int_0^1 \int_0^1 \frac{x^r y^r \log xy \, dx dy}{1 - xy} = 2\left(\zeta(3) - \left(1 + \frac{1}{2^3} + \cdots + \frac{1}{r^3}\right)\right).$$

Indeed, considering the integral

$$\int_0^1 \int_0^1 \frac{x^{r+t} y^{s+t} dx dy}{1 - xy}$$

we see that it is

$$\sum_{k=0}^{\infty} \frac{1}{(k + r + t + 1)(k + s + t + 1)}$$

$$= \frac{1}{r - s} \sum_{k=0}^{\infty} \left(\frac{1}{k + s + t + 1} - \frac{1}{k + r + t + 1}\right)$$

which telescopes to give the first part of the assertion fo $r > s$. If we differentiate with respect to t and set $t = 0$, we can deduce the second assertion. In case $r = s$, we put $t = 0$ to deduce the formula for $\zeta(2)$. If we differentiate with respect to t and set $t = 0$, we deduce the formula involving $\zeta(3)$. Beukers then looks at the integral

$$-\int_0^1 \int_0^1 \frac{\log xy}{1 - xy} P_n(x) P_n(y) dx dy$$

which by our observations above is

$$(C_n + D_n \zeta(3)) d_n^{-3}$$

with C_n, D_n integers. Since

$$-\frac{\log xy}{1 - xy} = \int_0^1 \frac{dz}{1 - (1 - xy)z},$$

the integral in question is

$$\int_0^1 \int_0^1 \int_0^1 \frac{P_n(x)P_n(y)dxdydz}{1-(1-xy)z}.$$

An n-fold integration by parts with respect to x gives that the integral is equal to

$$\int_0^1 \int_0^1 \int_0^1 \frac{(xyz)^n(1-x)^n P_n(y)dxdydz}{(1-(1-xy)z)^{n+1}}.$$

A change of variable $z = \frac{1-w}{1-(1-xy)w}$ followed by an n-fold integration by parts with respect to y gives that the integral is equal to

$$\int_0^1 \int_0^1 \int_0^1 \frac{(xyw)^n(1-x)^n(1-y)^n(1-w)^n}{(1-(1-xy)w)^{n+1}}dxdydw.$$

It turns out that the integral is bounded by

$$2(\sqrt{2}-1)^{4n}\zeta(3).$$

This can be deduced by noting that

$$\frac{(xyw)(1-x)(1-y)(1-w)}{(1-(1-xy)w)}$$

is bounded by $(\sqrt{2}-1)^4$ in the given region. Since the integral is non-zero, this with the other estimates for d_n gives the final result.

Picard–Fuchs Differential Equations and Modular Forms

Suppose we consider differential equations of the type

$$y^{(n)} + a_1(z)y^{(n-1)} + \cdots + a_n(z)y = 0$$

where the a_i are rational functions over the field of complex numbers (say) and y is a function of z. We would like to know when the equation admits n independent algebraic solutions. The complex numbers for which at least one of the rational functions a_i is not defined are called singular points of the differential equation and form a finite set S. At any non-singular point z_0, we may find a basis for the solution space of the differential equation at z_0. If we choose a closed path u beginning at z_0 contained in $\mathbb{P}^1\backslash S$ and analytically continue these solutions along this path, we find that when we return to z_0, we will have another basis of solutions. The change of basis matrix $\rho(u)$ depends only on the homotopy class of u and thus we may associate with each element of the fundamental group $\pi_1(\mathbb{P}^1\backslash S, z_0)$, an element of $GL(n, \mathbb{C})$. This defines a representation of $\pi_1(\mathbb{P}^1\backslash S, z_0)$ and is called the monodromy representation of the fundamental group. If we fix another non-singular point z_1, then the monodromy representation is conjugate to the earlier one. Thus the image of the monodromy representation is well defined in $GL(n, \mathbb{C})$ up to conjugacy and this we call the *monodromy group* of the differential equation.

In relation to singular points, one needs to make a distinction between a regular singular point and an irregular singular point. Given our differential equation, we say that a complex number z_0 is a *regular singular point* (see [54], for instance) if

$$\lim_{z \to z_0} (z - z_0)^i a_i(z)$$

exists for $i = 1, 2, \ldots, n$. The point ∞ is called a *regular singularity* if

$$\lim_{z \to \infty} z^i a_i(z)$$

exists and is finite for $i = 1, 2, \ldots, n$. The differential equation is called *Fuchsian* if every point of $\mathbb{P}^1$ is either non-singular or regular singular. It turns out that all the solutions of a Fuchsian equation are algebraic if and only if the monodromy group is finite.

It might be instructive to consider an example. Let us look at

$$z^2 y'' + \frac{1}{6} z y' + \frac{1}{6} y = 0.$$

It is readily verified that $z^{1/2}$ and $z^{1/3}$ are independent solutions of this equation. Observe that both solutions are algebraic. If we take the set of solutions $\{z^{1/2}, z^{1/3}\}$ which constitutes a basis and analytically continue this pair of solutions around a closed path containing zero, we get another basis of solutions $\{-z^{1/2}, e^{2\pi i/3} z^{1/3}\}$. The change of basis matrix is represented by

$$\begin{pmatrix} -1 & 0 \\ 0 & e^{2\pi i/3} \end{pmatrix}.$$

In this way, it is not difficult to see that the monodromy group is generated by this matrix which has order 6 and thus finite.

If we consider the equation

$$z^2 y'' - z y + y = 0$$

then the two independent solutions are $\{z, z \log z\}$, where the second solution is not algebraic. It is not hard to see that the monodromy is generated by the matrix

$$\begin{pmatrix} 1 & 2\pi i \\ 0 & 1 \end{pmatrix}$$

which is infinite cyclic.

An important class of Fuchsian equations is provided by the *hypergeometric differential equation* defined by

$$z(z - 1) y'' + [(a + b + 1)z - c] y' + ab y = 0$$

where a, b, c are real. The Euler–Gauss hypergeometric function

$$F(a, b, c; z) := \sum_{n=0}^{\infty} \frac{(a)_n (b)_n}{(c)_n n!} z^n$$

where $(x)_n = x(x+1)\cdots(x+n-1)$ is a solution. The points $0, 1$ and ∞ are regular singular points. In 1873, Schwarz determined the list of a, b, c for which the monodromy group is finite and this list is called Schwarz's list. For instance, $F(a, 1, 1; z) = (1-z)^{-a}$ is algebraic. The Chebyshev polynomials defined by $T_n(\cos z) = \cos nz$ are given by

$$F(-n, n, 1/2; (1-z)/2).$$

A similar formula exists for Legendre polynomials.

In our context, the Apéry recurrence relation can be translated into a differential equation. If we set

$$f(t) = \sum_{n=0}^{\infty} u_n t^n$$

then, we find the recurrence is equivalent to

$$(z^4 - 34z^3 + z^2)y''' + (6z^3 - 153z^2 + 3z)y'' + (7z^2 - 112z + 1)y' + (z - 5)y = 0.$$

It turns out that the solution space of this differential equation is spanned by the squares of a second-order equation which is

$$(t^3 - 34t^2 + t)y'' + (2t^2 - 51t + 1)y' + \frac{1}{4}(t - 10)y = 0.$$

We will now indicate briefly the "modular proof" of Beukers [20]. He begins with an elementary observation. Suppose that

$$f_0(t), f_1(t), \ldots, f_k(t)$$

are power series in t with rational coefficients. Suppose further that the n-th coefficient has denominator dividing $d^n [1, 2, \ldots, n]^r$ for some fixed d and r. Suppose that there are real numbers $\theta_1, \ldots, \theta_k$ such that

$$f_0(t) + \theta_1 f_1(t) + \cdots + \theta_k f_k(t)$$

has radius of convergence ρ and that infinitely many of its Taylor coefficients are non-zero. If $\rho > de^r$, then at least one of the $\theta_1, \ldots, \theta_k$ is irrational.

To prove this, write

$$f_i(t) = \sum_{n=0}^{\infty} a_{in} t^n.$$

Let $\epsilon > 0$, and choose n large so that

$$|a_{0n} + \theta_1 a_{1n} + \cdots + \theta_k a_{kn}| < (\rho - \epsilon)^{-n}.$$

If all the θ_i are rational, let D be the common denominator. Then,

$$S_n := Dd^n[1,2,\ldots,n]^r|a_{0n} + \theta_1 a_{1n} + \cdots + a_{kn}\theta_k|$$

is an integer smaller than

$$Dd^n[1,2,..,n]^r(\rho - \epsilon)^{-n}.$$

By the prime number theorem $[1,2,\ldots,n] < e^{(1+\epsilon)n}$ for large n and so by the hypothesis, we see that S_n vanishes for n sufficiently large.

With this general observation in mind, we consider the following.

Proposition 28.1 [20] *Let*

$$F(z) = \sum_{n=1}^{\infty} a_n q^n$$

be such that

$$F(-1/Nz) = w(-iz\sqrt{N})^k F(z),$$

where $w = \pm 1$. Let

$$f(z) = \sum_{n=1}^{\infty} \frac{a_n}{n^{k-1}} q^n,$$

and let

$$L(s, F) = \sum_{n=1}^{\infty} \frac{a_n}{n^s}.$$

Finally, set

$$h(z) = f(z) - \sum_{r=0}^{(k-3)/2} L(k-r-1, F)(2\pi i z)^r/r!.$$

Then

$$h(z) - D = (-1)^{k-1} w(-iz\sqrt{N})^{k-2} h(-1/Nz).$$

Here $D = 0$ if k is odd and otherwise equal to

$$(2\pi i z)^{k/2-1} \frac{L(k/2, F)}{(k/2 - 1)!}$$

if k is even. Further, $L(k/2, F) = 0$ if $w = -1$.

Proof. We apply a lemma of Hecke (as in [20]) to deduce that

$$f(z) - w(-1)^{k-1}(-iz\sqrt{N})^{k-2} f(-1/Nz) = \sum_{r=0}^{k-2} \frac{L(k-r-1, F)}{r!}(2\pi i z)^r.$$

Splitting the summation on the right-hand side into sums over $r < k/2 - 1$, $r > k/2 - 1$ and possibly $r = k/2 - 1$, and applying the functional equation

$$\frac{L(k-r-1,F)}{r!} = w(-1)^k(-i\sqrt{N})^{k-2}(-1/N)^{k-r-2}(2\pi i)^{k-2r-2}\frac{L(r+1,F)}{(k-r-2)!},$$

we obtain the result. $\square$

Note that the function f above is the *Eichler integral* associated with F. We apply this theorem to the following:

$$40F(z) = E_4(z) - 36E_4(6z) - 28E_4(2z) + 63E_4(3z)$$

and

$$24E(z) = -5E_2(z) + 30E_2(6z) + 2E_2(2z) - 3E_2(3z).$$

Here E_4 and E_2 are the usual Eisenstein series:

$$E_4(z) = 1 + 240 \sum_{n=1}^{\infty} \sigma_3(n)q^n, \qquad q = e^{2\pi i z},$$

$$E_2(z) = 1 - 24 \sum_{n=1}^{\infty} \sigma_1(n)q^n,$$

where

$$\sigma_k(n) = \sum_{d|n} d^k.$$

For a quick introduction to the notions of modular forms relevant here, we suggest the masterly article of Zagier in [25]. Let $\Gamma_1(6)$ be the subgroup of the full modular group $SL_2(\mathbb{Z})$ defined by

$$\left\{ \begin{pmatrix} a & b \\ c & d \end{pmatrix} : a,b,c,d \in \mathbb{Z}, ad-bc = 1, a \equiv d \equiv 1 \pmod 6, c \equiv 0 \pmod 6 \right\}.$$

One can show that F is a modular form of weight 4 on $\Gamma_1(6)$ and that

$$F(-1/6z) = -36z^4F(z)$$

and $F(i\infty) = 0$. Also E is a modular form of weight 2 on $\Gamma_1(6)$ and

$$E(-1/6z) = -6z^2E(z).$$

The Dirichlet series corresponding to $L(s,F)$ is

$$6(1 - 6^{2-s} - 7.2^{2-s} + 7.3^{2-s})\zeta(s)\zeta(s-3).$$

Define $f(z)$ by

$$(d/dz)^3 f(z) = (2\pi i)^3 F(z).$$

This is the Eichler integral associated with F. Further, $f(i\infty) = 0$. An application of the proposition gives

$$6z^2(f(-1/6z) - L(3, F)) = -(f(z) - L(3, F)).$$

Since $\zeta(0) = -1/2$, we have

$$L(3, F) = 6(-1/3)\zeta(3)\zeta(0) = \zeta(3)$$

and therefore

$$6z^2(f(-1/6z) - \zeta(3)) = -(f(z) - \zeta(3)).$$

Multiplication by $E(-1/6z) = -6z^2 E(z)$ gives

$$E(-1/6z)(f(-1/6z) - \zeta(3)) = E(z)(f(z) - \zeta(3)).$$

The field of modular functions for the group $\Gamma_1(6)$ is generated by

$$t(z) = \left(\frac{\Delta(6z)\Delta(z)}{\Delta(3z)\Delta(2z)}\right)^{1/2} = q \prod_{n=0}^{\infty}(1 - q^{6n+1})^{12}(1 - q^{6n+5})^{-12},$$

where

$$\Delta(z) = q \prod_{n=1}^{\infty}(1 - q^n)^{24}$$

is Ramanujan's cusp form of weight 12 for the full modular group. From this, we see that

$$q = t + 12t^2 + 222t^3 + \cdots.$$

Also,

$$E(z) = 1 + 5q + 13q^2 + \cdots$$

and hence

$$E(t) = 1 + 5t + 73t^2 + 1445t^3 + \cdots.$$

Similarly,

$$E(t)f(t) = 6t + (351/4)t^2 + (62531/36)t^3 \cdots.$$

By construction, one notes that $E(t)$ has integral coefficients and that

$$E(t)f(t) = \sum_{n=0}^{\infty} a_n t^n$$

has coefficients which are rational and $a_n[1, 2, \ldots, n]^3$ are integers.

Here we are working with the power series $E(t)f(t)$ and $E(t)$. By appealing to the earlier observation of Beukers about power series with rational coefficients and exploiting the identity

$$E(-1/6z)(f(-1/6z) - \zeta(3)) = E(z)(f(z) - \zeta(3)),$$

we deduce the irrationality of $\zeta(3)$.

It is interesting to note that the coefficients of these two power series are precisely the Apéry numbers and the recurrence relations now become irrelevant. However, the recurrence relations can be derived from the following general principle of expressing modular forms as solutions of linear differential equations.

If $F(z)$ is a modular form of weight k and $t(z)$ is a modular function, then $F(t)$ (locally) satisfies a differential equation of order $k + 1$. This seems to be a "folklore" theorem. There is a memoir of Stiller [120] that discusses this theorem in some detail. We provide a short summary of the ideas involved.

Suppose for the sake of simplicity, we have a modular form f of weight 2 for some subgroup Γ of $SL_2(\mathbf{Z})$. If we consider the three-dimensional vector space spanned by $f(z), zf(z)$ and $z^2 f(z)$, then it is easy to see that $\gamma \in \Gamma$ acts on this vector space in the obvious way. For example,

$$\gamma \cdot f(z) = f\left(\frac{az + b}{cz + d}\right) = (cz + d)^2 f(z) = c^2 z^2 f(z) + 2cdz f(z) + d^2 f(z),$$

and it is not difficult to see that

$$\gamma \cdot \begin{pmatrix} z^2 f(z) \\ z f(z) \\ f(z) \end{pmatrix} = \begin{pmatrix} a^2 & 2ab & b^2 \\ ac & ad + bc & bd \\ c^2 & 2cd & d^2 \end{pmatrix} \begin{pmatrix} z^2 f(z) \\ z f(z) \\ f(z) \end{pmatrix}.$$

If we denote the matrix on the right-hand side of the equation by

$$M = \mathrm{Sym}^2(\gamma),$$

then a direct verification shows that it is of determinant one. Writing $F(z) = (z^2 f(z), z f(z), f(z))$, the above formula reads as

$$\gamma.F(z) = MF(z).$$

Now let us consider f as a function of t. Then, one verifies that

$$\frac{d}{dt}F(\gamma z) = M\frac{d}{dt}F(z)$$

by checking it for each of the entries, noting that t is Γ-invariant. Further, the same is true for the higher t-derivatives. If we are trying to find the differential equation satisfied by three functions f_2, f_1, f_0 say, then we begin by assuming it is of the form

$$y''' + a_2(t)y'' + a_1(t)y' + a_0(t)y = 0.$$

If we let $f_0(t) = f(t), f_1(t) = tf(t), f_2(t) = t^2 f(t)$ and would like to determine the differential equation these functions satisfy, then by Cramer's rule, we can write down $a_0(t), a_1(t), a_2(t)$ in the obvious way using determinants. For instance, one of the coefficients is given by the determinant

$$\begin{vmatrix} f_0' & f_0'' & f_0''' \\ f_1' & f_1'' & f_1''' \\ f_2' & f_2'' & f_2''' \end{vmatrix}.$$

Thus we do obtain a differential equation of order 3. This differential equation has coefficients in terms of our solutions f_0, f_1, f_2. The action of γ on them is the same as multiplying by the symmetric square matrix M of γ having determinant one. Thus the coefficients are Γ-invariant meromorphic functions and therefore must be algebraic functions of t. Hence we are done. This equation is called the Picard–Fuchs differential equation associated with $X(\Gamma)$.

In our case, we take the weight 2 form

$$g(z) = \frac{\eta(2z)^7 \eta(3z)^7}{\eta(z)^5 \eta(6z)^5}$$

where η denotes the Dedekind η-function. Then g is a modular form of weight 2 for $\Gamma_1(6)$ and is equal to the function E in Beukers' proof. Again, let t be the function

$$t(z) = \left(\frac{\Delta(6z)\Delta(z)}{\Delta(3z)\Delta(2z)} \right)^{1/2} = q \prod_{n=0}^{\infty} (1 - q^{6n+1})^{12}(1 - q^{6n+5})^{-12}.$$

Treating g as a function of t, we can derive a third-order differential equation. From this differential equation, we can recover the recursion defining the integral Apéry numbers. The integrality is a consequence of the observation that both $g(z)$ and $t(z)$ have integral q-expansions and that $t(z)$ is normalised (i.e. starts with q).

On the other hand, the recursion and divisibility properties of the rational Apéry numbers are more involved as they are related to the function F introduced before in Beukers' proof. This function is defined as

$$40F(z) = E_4(z) - 36E_4(6z) - 28E_4(2z) + 63E_4(3z),$$

and which being a modular form of weight 4, satisfies a differential equation of order 5.

However as suggested by Beukers himself, it is more convenient to work with the function fg where f is the Eichler integral associated with F. Expressing fg as a function of t and working with the associated differential equation, we can recover the recurrence formula for the rational Apéry numbers as well as the divisibility properties enjoyed by their denominators. We recommend the article of Zagier in [25] for a more elaborate account.

The theme of expressing modular forms as functions of modular functions and thereby realising them as solutions of linear differential equations of finite order constitutes a venerable theme. The imprints of this can be traced in the works of past masters like Gauss, Fricke, Klein, Poincare, Ramanujan, etc. As we have seen before, the fact that the $\mathbb{C}$-algebra generated by the Eisenstein series E_2, E_4 and E_6 is closed under differentiation constitutes an essential ingredient in the work of Nesterenko.

Finally, it is not clear if any of these proofs of irrationality of $\zeta(3)$ can yield irrationality of other odd zeta values like that of $\zeta(5)$. However, we do have the following theorem of Rivoal [14, 105] which is the most general result in this context.

Theorem 28.2 *Given any $\epsilon > 0$, there exists an integer $N = N(\epsilon)$ such that for all $n > N$, the dimension of the $\mathbb{Q}$-vector space generated by the numbers*

$$1, \; \zeta(3), \ldots, \zeta(2n-1), \; \zeta(2n+1)$$

exceeds

$$\frac{1-\epsilon}{1+\log 2}\log n.$$

In particular, Rivoal proved that infinitely many odd zeta values are irrational. Concerning the individual odd zeta values, Rivoal [106] himself showed that at least one of the nine numbers

$$\zeta(5), \;\; \zeta(7), \;\; \ldots \;\; ,\zeta(21)$$

is irrational. This was sharpened by Zudilin [137] who showed that at least one among the four numbers

$$\zeta(5), \;\; \zeta(7), \;\; \zeta(9), \;\; \zeta(11)$$

is irrational. Thus the irrationality, let alone transcendence, of odd zeta values seems to be a very hard question.

Finally, in this mysterious modular-transcendence conundrom, one can ask about nature of the values taken by the L-functions associated with modular forms. Indeed, Kohnen [71] (see also [60, 72]) has made general conjectures regarding special values of L-series attached to modular forms of weight $2k$ for the full modular group. These conjectures when applied to classical Eisenstein series imply the transcendence of $\zeta(2k+1)/\pi^{2k+1}$ for all $k \geq 1$. But this is a different journey which we do not undertake here.

Bibliography

[1] W.W. Adams, On the algebraic independence of certain Liouville numbers. J. Pure Appl. Algebra **13**(1), 41–47 (1978)

[2] S.D. Adhikari, N. Saradha, T.N. Shorey, R. Tijdeman, Transcendental infinite sums. Indag. Math. (N.S.) **12**(1), 1–14 (2001)

[3] R. Apéry, Irrationalité de $\zeta(2)$ et $\zeta(3)$. Astérisque **61**, 11–13 (1979)

[4] T. Apostol, *Introduction to Analytic Number Theory*. Undergraduate Texts in Mathematics (Springer, Berlin, 1976)

[5] J. Ax, On the units of an algebraic number field. Ill. J. Math. **9**, 584–589 (1965)

[6] J. Ax, On Schanuel's conjectures. Ann. Math. **93**(2), 252–268 (1971)

[7] J. Ayoub, Une version relative de la conjecture des périodes de Kontsevich–Zagier. Ann. Math. (to appear)

[8] A. Baker, Linear forms in the logarithms of algebraic numbers. Mathematika **13**, 204–216 (1966)

[9] A. Baker, *Transcendental Number Theory* (Cambridge University Press, Cambridge, 1975)

[10] A. Baker, B. Birch, E. Wirsing, On a problem of Chowla. J. Number Theory **5**, 224–236 (1973)

[11] A. Baker, G. Wüstholz, Logarithmic forms and group varieties. J. Reine Angew. Math. **442**, 19–62 (1993)

[12] A. Baker, G. Wüstholz, *Logarithmic Forms and Diophantine Geometry*. New Mathematical Monographs (Cambridge University Press, Cambridge, 2007)

[13] R. Balasubramanian, S. Gun, On zeros of quasi-modular forms. J. Number Theory **132**(10), 2228–2241 (2012)

[14] K. Ball, T. Rivoal, Irrationalité dune infinité de valeurs de la fonction zeta aux entiers impairs. Invent. Math. **146**(1), 193–207 (2001)

[15] K. Barré-Sirieix, G. Diaz, F. Gramain, G. Philibert, Une preuve de la conjecture de Mahler-Manin. Invent. Math. **124**, 1–9 (1996)

[16] C. Bertolin, Périodes de 1-motifs et transcendence. J. Number Theory **97**(2), 204–221 (2002)

[17] D. Bertrand, Séries d'Eisenstein et transcendance. Bull. Soc. Math. France **104**(3), 309–321 (1976)

[18] D. Bertrand, D. Masser, Linear forms in elliptic integrals. Invent. Math. **58**, 283–288 (1980)

[19] F. Beukers, A note on the irrationality of $\zeta(2)$ and $\zeta(3)$. Bull. Lond. Math. Soc. **11**, 268–272 (1979)

[20] F. Beukers, Irrationality proofs using modular forms. Astérisque **147/148**, 271–283 (1987)

[21] F. Brown, Mixed Tate motives over $\mathbb{Z}$. Ann. Math. **175**(2), 949–976 (2012)

[22] W.D. Brownawell, The algebraic independence of certain numbers related by the exponential function. J. Number Theory **6**, 22–31 (1974)

[23] W.D. Brownawell, K.K. Kubota, The algebraic independence of Weierstrass functions and some related numbers. Acta Arith. **33**(2), 111–149 (1977)

[24] A. Brumer, On the units of algebraic number fields. Mathematika **14**, 121–124 (1967)

[25] J. Bruinier, G. van der Geer, G. Harder, D. Zagier, *The 1-2-3 of Modular Forms*. Universitext (Springer, Berlin, 2008)

[26] P. Bundschuh, Zwei Bemerkungen über transzendente Zahlen. Monatsh. Math. **88**(4), 293–304 (1979)

[27] P. Cartier, Fonctions polylogarithmes, nombres polyzetâs et groupes pro-unipotents, in *Séminaire Bourbaki*, vol. 2000/2001, Astérisque No. 282, Exp. No. 885 (2002), pp. 137–173

[28] K. Chandrasekharan, *Elliptic Functions*. Grundlehren der Mathematischen Wissenschaften, vol. 281 (Springer, Berlin, 1985)

[29] S. Chowla, A special infinite series. Norske Vid. Selsk. Forth. (Trondheim) **37**, 85–87 (1964) (see also Collected Papers, vol. 3, pp. 1048–1050)

[30] S. Chowla, A. Selberg, On Epstein's zeta-function. J. Reine Angew. Math. **227**, 86–110 (1967)

[31] S. Chowla, The nonexistence of nontrivial linear relations between roots of a certain irreducible equation. J. Number Theory **2**, 120–123 (1970)

[32] G.V. Chudnovsky, Algebraic independence of constants connected with the exponential and the elliptic functions. Dokl. Akad. Nauk Ukrain. SSR Ser. A **8**, 698–701 (1976)

[33] G.V. Chudnovsky, *Contributions to the Theory of Transcendental Numbers*. Mathematical Surveys and Monographs, vol. 19 (American Mathematical Society, Providence, 1974)

[34] R.F. Coleman, On a stronger version of the Schanuel–Ax theorem. Am. J. Math. **102**(4), 595–624 (1980)

[35] P. Colmez, Résidu en s=1 des fonctions zeta p-adiques. Invent. Math. **91**(2), 371–389 (1988)

[36] D. Cox, *Primes of the Form $x^2 + ny^2$* (Wiley, New York, 1989)

[37] H. Davenport, *Multiplicative Number Theory*, vol. 74, 2nd edn. (Springer, New York, 1980)

[38] P. Deligne, Valeurs de fonctions L et periodes d'integrales. Proc. Symp. Pure Math. **33**(2), 313–346 (1979)

[39] P. Deligne, J.S. Milne, A. Ogus, K. Shih, *Hodge Cycles, Motives, and Shimura Varieties*. Lecture Notes in Mathematics, vol. 900 (Springer, Berlin, 1982)

[40] P. Deligne, A. Goncharov, Groupes fondamentaux monitrices de Tate mixte. Ann. Sci. Ec. Norm. Sup. **38**(4), 1–56 (2005)

[41] P. Deligne, Multizêtas, d'aprés Francis Brown, in *Séminaire Bourbaki*, Exp. 1048 (2011–2012)

[42] F. Diamond, J. Shurman, *A First Course in Modular Forms*. Graduate Texts in Mathematics, vol. 228 (Springer, New York, 2005)

[43] G. Diaz, Grands degrés de transcendance pour des familles d'exponentielles. C.R. Acad. Sci. Paris. Sér. I Math. **305**(5), 159–162 (1987)

[44] G. Diaz, La conjecture des quatre exponentielles et les conjectures de D. Bertrand sur la fonction modulaire. J. Théor. Nombres Bord. **9**(1), 229–245 (1997)

[45] A. El Basraoui, A. Sebbar, Zeros of the Eisenstein series E_2. Proc. Am. Math. Soc. **138**(7), 2289–2299 (2010)

[46] E. Freitag, R. Busam, *Complex Analysis*. Universitext (Springer, Berlin, 2009)

[47] A.O. Gel'fond, Sur le septième problème de Hilbert. Izv. Akad. Nauk. SSSR **7**, 623–630 (1934)

[48] A.O. Gel'fond, On algebraic independence of algebraic powers of algebraic numbers. Dokl. Akad. Nauk. SSSR **64**, 277–280 (1949)

[49] A. Ghosh, P. Sarnak, Real zeros of holomorphic Hecke cusp forms. J. Eur. Math. Soc. **14**(2), 465–487 (2012)

[50] D. Goldfeld, The class number of quadratic fields and the conjectures of Birch and Swinnerton-Dyer. Ann. Scuola Norm. Sup. Pisa Cl. Sci. **3**(4), 624–663 (1976)

[51] D. Goldfeld, The conjectures of Birch and Swinnerton-Dyer and the class numbers of quadratic fields. Astérisque **41–42**, 219–227 (1977)

[52] D. Goldfeld, Gauss's class number problem for imaginary quadratic fields. Bull. Am. Math. Soc. (N.S.) **13**(1), 23–37 (1985)

[53] P. Grinspan, Measures of simultaneous approximation for quasi-periods of abelian varieties. J. Number Theory **94**(1), 136–176 (2002)

[54] P.A. Griffiths, Periods of integrals on algebraic manifolds: summary of main results and discussion of open problems. Bull. Am. Math. Soc. **76**, 228–296 (1970)

[55] B.H. Gross, On an identity of Chowla and Selberg. J. Number Theory **11**, 344–348 (1979)

[56] B. Gross, D. Zagier, Heegner points and derivatives of L-series. Invent. Math. **84**(2), 225–320 (1986)

[57] S. Gun, Transcendental zeros of certain modular forms. Int. J. Number Theory **2**(4), 549–553 (2006)

[58] S. Gun, M. Ram Murty, P. Rath, Transcendental nature of special values of L-functions. Can. J. Math. **63**, 136–152 (2011)

[59] S. Gun, M. Ram Murty, P. Rath, Algebraic independence of values of modular forms. Int. J. Number Theory **7**(4), 1065–1074 (2011)

[60] S. Gun, M. Ram Murty, P. Rath, Transcendental values of certain Eichler integrals. Bull. Lond. Math. Soc. **43**(5), 939–952 (2011)

[61] S. Gun, M. Ram Murty, P. Rath, On a conjecture of Chowla and Milnor. Can. J. Math. **63**(6), 1328–1344 (2011)

[62] S. Gun, M. Ram Murty, P. Rath, A note on special values of L-functions. Proc. Am. Math. Soc. **142**(4), 1147–1156 (2014)

[63] S. Gun, M. Ram Murty, P. Rath, Linear independence of Hurwitz zeta values and a theorem of Baker-Birch-Wirsing over number fields. Acta Arith. **155**(3), 297–309 (2012)

[64] R.C. Gunning, H. Rossi, *Analytic Functions of Several Complex Variables* (AMS Chelsea, Providence, 2009)

[65] R. Holowinsky, Sieving for mass equidistribution. Ann. Math. **172**(2), 1499–1516 (2010)

[66] R. Holowinsky, K. Soundararajan, Mass equidistribution for Hecke eigenforms. Ann. Math. **172**(2), 1517–1528 (2010)

[67] D. Husemoller, *Elliptic Curves.* Graduate Texts in Mathematics, vol. 111 (Springer, Berlin, 2004)

[68] Y. Ihara, V. Kumar Murty, M. Shimura, On the logarithmic derivatives of Dirichlet *L*-functions at $s = 1$. Acta Arith. **137**(3), 253–276 (2009)

[69] N. Kanou, Transcendency of zeros of Eisenstein series. Proc. Jpn. Acad. Ser. A Math. Sci. **76**(5), 51–54 (2000)

[70] N. Koblitz, *Introduction to Elliptic Curves and Modular Forms.* Graduate Texts in Mathematics, vol. 97 (Springer, Berlin, 1993)

[71] W. Kohnen, Transcendence conjectures about periods of modular forms and rational structures on spaces of modular forms. Proc. Indian Acad. Sci. (Math. Sci.) **99**(3), 231–233 (1989)

[72] W. Kohnen, D. Zagier, Modular forms with rational periods, in *Modular Forms*, ed. by R. Rankin. Ellis Horwood Series in Mathematics and Its Applications. Statistics and Operational Research (Horwood, Chichester, 1984), pp. 197–249

[73] W. Kohnen, Transcendence of zeros of Eisenstein series and other modular functions. Comment. Math. Univ. St. Pauli **52**(1), 55–57 (2003)

[74] M. Kontsevich, D. Zagier, Periods, in *Mathematics Unlimited-2001 and Beyond* (Springer, Berlin, 2001), pp. 771–808

[75] S. Lang, With appendixes by D. Zagier and Walter Feit, in *Introduction to Modular Forms*, vol. 222 (Springer, Berlin, 1976) (Corrected reprint of the 1976 original)

[76] S. Lang, *Algebraic Number Theory.* Graduate Texts in Mathematics, vol. 110 (Springer, Berlin, 2000)

[77] S. Lang, *Elliptic Functions.* Graduate Texts in Mathematics, vol. 112, 2nd edn. (Springer, Berlin, 1987)

[78] S. Lang, Algebraic values of Meromorphic functions I. Topology **3**, 183–191 (1965)

[79] S. Lang, *Introduction to Transcendental Numbers*. (Addison-Wesley, Reading, 1966)

[80] F. Lindemann, Über die zahl π, Math. Annalen, **20**, 213–225 (1882)

[81] K. Mahler, Remarks on a paper by W. Schwarz. J. Number Theory **1**, 512–521 (1969)

[82] Y. Manin, Cyclotomic fields and modular curves. Russ. Math. Surv. **26**(6), 7–78 (1971)

[83] D. Masser, *Elliptic Functions and Transcendence*. Lecture Notes in Mathematics, vol. 437 (Springer, Berlin, 1975)

[84] Y.V. Nesterenko, Modular functions and transcendence. Math. Sb. **187**(9), 65–96 (1996)

[85] Y.V. Nesterenko, Some remarks on $\zeta(3)$. Math. Notes **59**, 625–636 (1996)

[86] Y.V. Nesterenko, P. Philippon (eds.), *Introduction to Algebraic Independence Theory*. Lecture Notes in Mathematics, vol. 1752 (Springer, Berlin, 2001)

[87] Y.V. Nesterenko, Algebraic independence for values of Ramanujan functions, in *Introduction to Algebraic Independence Theory*, ed. by Y.V. Nesterenko, P. Philippon. Lecture Notes in Mathematics, vol. 1752 (Springer, 2001), pp. 27–46

[88] J. Neukirch, *Algebraic Number Theory*, vol. 322 (Springer, Berlin, 1999)

[89] J. Neukirch, A. Schmidt, K. Winberg, *Cohomology of Number Fields* (Springer, Berlin, 2000)

[90] J. Oesterlé, Nombres de classes des corps quadratiques imaginaires, in *Seminar Bourbaki*, vol. 1983/84, Astérisque No. 121–122 (1985), pp. 309–323

[91] P. Philippon, Variétés abéliennes et indépandance algébrique. II. Un analogue abélien du théoréme de Lindemann-Weierstrass. Invent. Math. **72**(3), 389–405 (1983)

[92] P. Philippon, Critères pour l'independance algébrique. Inst. Hautes Études Sci. Publ. Math. **64**, 5–52 (1986)

[93] G. Prasad, A. Rapinchuk, Weakly commensurable arithmetic groups and isospectral locally symmetric spaces. Inst. Hautes Études Sci. Publ. Math. **109**, 113–184 (2009)

[94] M. Ram Murty, *Problems in Analytic Number Theory*. Graduate Texts in Mathematics, Readings in Mathematics, vol. 206, 2nd edn. (Springer, New York, 2008)

[95] M. Ram Murty, An introduction to Artin L-functions. J. Ramanujan Math. Soc. **16**(3), 261–307 (2001)

[96] M. Ram Murty, V. Kumar Murty, Transcendental values of class group L-functions. Math. Ann. **351**(4), 835–855 (2011)

[97] M. Ram Murty, V. Kumar Murty, A problem of Chowla revisited. J. Number Theory **131**(9), 1723–1733 (2011)

[98] M. Ram Murty, V. Kumar Murty, Transcendental values of class group L-functions-II. Proc. Am. Math. Soc. **140**(9), 3041–3047 (2012)

[99] M. Ram Murty, N. Saradha, Transcendental values of the digamma function. J. Number Theory **125**(2), 298–318 (2007)

[100] M. Ram Murty, C.J. Weatherby, On the transcendence of certain infinite series. Int. J. Number Theory **7**(2), 323–339 (2011)

[101] M. Ram Murty, Some remarks on a problem of Chowla. Ann. Sci. Math. Qué. **35**(2), 229–237 (2011)

[102] K. Ramachandra, Some applications of Kronecker's limit formulas. Ann. Math. **80**(2), 104–148 (1964)

[103] K. Ramachandra, On the units of cyclotomic fields. Acta Arith. **12**, 165–173 (1966/1967)

[104] F.K.C. Rankin, H.P.F. Swinnerton-Dyer, On the zeros of Eisenstein series. Bull. Lond. Math. Soc. **2**, 169–170 (1970)

[105] T. Rivoal, La fonction zeta de Riemann prend une infinité de valeurs irrationnelles aux entiers impairs. C. R. Acad. Sci. Paris Sér. I Math. **331**(4), 267–270 (2000)

[106] T. Rivoal, Irrationalité d'au moins un des neuf nombres $\zeta(5), \zeta(7), \cdots, \zeta(21)$. Acta Arith. **103**(2), 157–167 (2002)

[107] D. Roy, An arithmetic criterion for the values of the exponential function. Acta Arith. **97**(2), 183–194 (2001)

[108] Z. Rudnick, On the asymptotic distribution of zeros of modular forms. Int. Math. Res. Notes **34**, 2059–2074 (2005)

[109] J.-P. Serre, *A Course in Arithmetic*, vol. 7 (Springer, Berlin, 1973)

[110] T. Schneider, Arithmetische Untersuchungen elliptischer Integrale. Math. Ann. **113**(1), 1–13 (1937)

[111] E. Scourfield, On ideals free of large prime factors. J. Théor. Nombres Bord. **16**(3), 733–772 (2004)

[112] G. Shimura, *An Introduction to the Arithmetic Theory of Automorphic Functions* (Princeton University Press, Princeton, 1994)

[113] C.L. Siegel, *Advanced Analytic Number Theory*. Tata Institute of Fundamental Research Studies in Mathematics, vol. 9, 2nd edn. (Tata Institute of Fundamental Research, Bombay, 1980)

[114] J. Silverman, *The Arithmetic of Elliptic Curves*. Graduate Texts in Mathematics, vol. 106, 2nd edn. (Springer, New York, 1986)

[115] J. Silverman, *Advanced Topics in the Arithmetic of Elliptic Curves*. Graduate Texts in Mathematics, vol. 151 (Springer, New York, 1994)

[116] H.M. Stark, On complex quadratic fields with class number equal to one. Trans. Am. Math. Soc. **122**, 112–119 (1966)

[117] H.M. Stark, A complete determination of the complex quadratic fields of class-number one. Mich. Math. J. **14**, 1–27 (1967)

[118] H.M. Stark, L-functions at $s = 1$. II. Artin L-functions with rational characters. Adv. Math. **17**(1), 60–92 (1975)

[119] K. Soundararajan, Quantum unique ergodicity for $SL_2(\mathbb{Z}) \backslash \mathbb{H}$. Ann. Math. **172**(2), 1529–1538 (2010)

[120] P. Stiller, Special values of Dirichlet series, monodromy and the periods of automorphic forms. Mem. Am. Math. Soc. **49**(299) (1984)

[121] T. Terasoma, Mixed Tate motives and multiple zeta values. Invent. Math. **149**(2), 339–369 (2002)

[122] A. Van der Poorten, On the arithmetic nature of definite integrals of rational functions. Proc. Am. Math. Soc. **29**, 451–456 (1971)

[123] A. Van der Poorten, A proof that Euler missed ... Apéry's proof of the irrationality of $\zeta(3)$. Math. Intell. **1**(4), 195–203 (1978/1979)

[124] K.G. Vasilev, On the algebraic independence of the periods of abelian integrals. Mat. Zametki **60**(5), 681–691 (1996)

[125] M. Waldschmidt, *Diophantine Approximation on Linear Algebraic Groups*, vol. 326 (Springer, Berlin, 2000)

[126] M. Waldschmidt, Transcendence of periods: the state of the art. Pure Appl. Math. Q. **2**(2), 435–463 (2006)

[127] M. Waldschmidt, Elliptic functions and transcendence, in *Surveys in Number Theory*. Developments in Mathematics, vol. 17 (Springer, 2008), pp. 143–188

[128] M. Waldschmidt, Solution du huitième problème de Schneider. J. Number Theory **5**, 191–202 (1973)

[129] M. Waldschmidt, Transcendance et exponentielles en plusieurs variables. Invent. Math. **63**(1), 97–127 (1981)

[130] L. Washington, *Introduction to Cyclotomic Fields*, vol. 83 (Springer, Berlin, 1997)

[131] A. Weil, *Elliptic Functions According to Eisenstein and Kronecker* (Springer, Berlin, 1999)

[132] E.T. Whittaker, G.N. Watson, *A Course of Modern Analysis*, 4th edn. (Cambridge University Press, Cambridge, 1927)

[133] G. Wüstholz, Über das Abelsche Analogon des Lindemannschen Satzes. Invent. Math. **72**(3), 363–388 (1983)

[134] D. Zagier, Values of zeta functions and their applications, in *First European Congress of Mathematics*, vol. 2 (1992), pp. 497–512

[135] D. Zagier, Evaluation of the multiple zeta values $\zeta(2, \cdots, 2, 3, 2, \cdots, 2)$. Ann. Math. **175**(2), 977–1000 (2012)

[136] B. Zilber, Exponential sums equations and the Schanuel conjecture. J. Lond. Math. Soc. 2 **65**(1), 27–44 (2002)

[137] V.V. Zudilin, One of the numbers $\zeta(5), \zeta(7), \zeta(9), \zeta(11)$ is irrational. Uspekhi Mat. Nauk **56**(4(340)), 149–150 (2001) (translation in Russ. Math. Surv. **56**(4), 774–776, 2001)

Index